Make Your Factory Great & Keep It That Way

By
Liam Cassidy

A serial fixer of factories on three continents shares his how-to secrets for sustainable continuous improvement

Foreword by Chet Marchwinski

ISBN: 978-1-3999-1802-2

All photos courtesy of the author. Reasonable efforts have been made to contact copyright holders for permission and give credit. If additional information is provided, changes may be made to future printings.

This book reflects the author's present recollections of experiences over time, and all content within regarding companies, persons, and experiences is strictly his opinion. Some names and characteristics have been changed, some events have been compressed, and some dialogue has been recreated.

Editing by Dustin Bilyk @ www.authorshand.com
Cover Design by Aoife Cassidy

Printed by Ingram Spark in the United States of America. First printing edition 2022.

Recommendations

I'm very proud and honored to write this endorsement for my good friend Liam Cassidy.

In 2010, I met Liam in China when my boss sent me a consultant to help us with Lean Management Implementation. At that time, I was leading a thousand-person operation with a high vertical integration level. The site in Jinan Shandong had to be integrated into a high-performance global organisation.

It was probably one of the busiest and most challenging times in my life. Liam and I developed a strong and trusted relationship, which turned into a long friendship. Not only was Liam a great advisor in Lean Management matters, but he also became a real coach for me in so many aspects of a top management career. His understanding of people, leadership, and world class manufacturing operations is unique, and he taught me a lot about complex and cultural change management.

The business transformation in China was exceptional and, over five years, we developed a strong and sustainable, high-performance organisation with great people at the top. After that, Liam and I worked together on many additional and successful initiatives. Just over a year ago, Altix acquired Liam's consultancy group. I was delighted to welcome him and his team into our organisation.

This book will let you into many of his secrets and will allow you, the reader, to ensure you too can have a great factory!

Yannick Schilly – Founding Partner, President & CEO
Altix Consulting

Make Your Factory Great & Keep it That Way is Liam Cassidy's account of leading manufacturing operations on a Journey to Excellence. Liam tells the stories of the operations in which he had direct responsibility, yet his story and influence extends much further than that.

I had the pleasure and good fortune to work with Liam at Gillette. Our global operations included 37 factories and packaging centres, which produced and supplied Gillette, Oral-B, Braun, and Duracell products. Liam's knowledge, determination, and vision impacted our entire manufacturing world. In the first eight years of our journey to Leaner operations, we offset the cost of inflation and then reduced manufacturing costs by 3 to 5% *every year*. Better still, we improved every one of our key performance metrics, and most importantly, that of employee engagement and satisfaction.

The key to this journey, for me, was first realising just how much untapped potential we had throughout our organisation, and then trusting experts, like Liam, to guide us toward achieving excellence in everything we did.

Tomas Lyden, Vice President Global Operations
Gillette Corporation (Retired)

It was my good fortune to serve four years on the Oral-B Iowa City Leadership Team during Liam Cassidy's tenure as Site Manager. In my 31 years with Oral-B/P&G, those were easily the most challenging and rewarding years of my career. I had a front row seat in seeing an organisation transformed from not knowing what we didn't know to one of great confidence, flexibility, and reliability.

Liam quickly established a new sense of purpose throughout the site. It was difficult at first to accept that the largest toothbrush manufacturing facility in the world could be relocated. He didn't sugarcoat the message. *Change needed to happen.* He quickly rolled out an aggressive Lean Program that involved everyone on site, and he transformed us into the best performing factory in the organisation.

I will always be grateful to him for allowing me to be a part of the process.

Bill Anciaux, Supply Chain Director
Oral-B Laboratories, Iowa City (Retired)

Liam is the best coach I ever met in my career life. I learnt a lot from him personally, and it was my luck to be hired by him and later work for him at both Braun and Festo. He has a magic key to Lean Deployment and knows how to activate a person, a team, and eventually an entire plant to make it happen. He is always full of energy and is a role model on how to drive Lean deployment. Whenever he arrived at the plant, he observed the ground activities by himself and said hello to everyone he met on the plant tour, giving his personal advice to the activities he just passed by. He is now disclosing all his magic to us in this great book. Enjoy and have fun!

Xu Xiaoying (Xxy) Operations Director, Asia Pacific,
Connecting Systems APTIV

I've known Liam Cassidy for almost a quarter of a century. Our cooperation began when we both ran Gillette factories in Ireland in the late 90's. First, we were peers, and later, Liam worked for me, running the Irish Gillette factories while I was in charge of all Braun and Oral-B factories globally. Liam always impressed me with his dedication, his consequent application of Lean principles, and his ability to excite and take with him the entire factory workforce. The results he delivered always surpassed expectations by far.

It was therefore no surprise that I hired him as a consultant once I left Gillette and moved on to Festo as Head of Global Production. At Festo, we had some traditionally organised factories that had not implemented Lean, so we had Liam support plants in Czech Republic, Hungary, USA, Germany, and India with great success.

Festo had acquired a state-owned pneumatic company in Jinan, China as well. Conditions and work standards were appalling, and with over 1000 employees, we were very concerned about the challenges there. But within nine months, Liam and his team had a Lean Program rolled out, and a strong sense of urgency existed throughout the site. Within three years, it became the best performing site in our organisation.

Liam and his team are true Lean experts; the best I've ever worked with in my long career as head of factories. This book is an excellent guide to demonstrate how factories can become great!

Christian Leonhard, Partner, Altix Consulting Inc.
Senior Vice President Global Operations Festo (Retired)

It was such an amazing journey working with Liam Cassidy while he led the Braun plant in Shanghai. It was a great honour to be on his team and a witness to the shining transformation.

Liam built up a strong team and created a solid base for Braun Shanghai to shift from a Braun factory to a Proctor & Gamble device centre serving Braun, Oral-B, and Gillette by beating the outsource manufacturing suppliers with better service and lower cost. This made it possible to keep increasing efficiency without reducing our workforce.

During the process, we witnessed the power of caring for people by engaging, envisioning, enabling, energising, and executing. We witnessed the team attitude change from nonchalant to proactive. We saw the effectiveness/efficiency difference between forced and motivated.

I will always be grateful to Liam for leading me through those exciting years, which later benefitted my whole career.

This book is a great guide for those wishing to create a great factory.

Rex Wang, Lean Manufacturing Director
OU AP, TK Elevator

It was my great pleasure to work for Liam as Head of Production in the Braun & Oral-B Plant in Shanghai. He is a true Lean expert with passion, and he is a great leader who drives change.

Implementing a Lean Process in isolation has very limited benefits and will not sustain itself long term. The management team in our plant jointly worked out a common vision about the factory and started the journey to excellence. In just half a year, our plant significantly improved its material flow, visual management on shopfloor, and upgraded to a high level of 5S standards. More important change happened: employees started to be engaged in our journey because they felt respect and recognition when they saw the improvement in our cloakroom and canteen service. Continuous Improvement and total involvement started to take root in employees' hearts.

Liam is one of two great men who influenced me the most in my career. "Lean can be only achieved by showing leadership on shopfloor with engagement of entire organisation." I take this as my belief. Several years later, I followed his recommendation and joined Festo where we started two years of cooperation again. No surprise, we repeated the success by engaging employees and changing the mindset of the organisation. Now, the factory in Jinan is the beacon of Festo global factories: best QCD (Quality, Cost, Delivery) result, a high level of automation, and a mature Lean production system.

I'm now heading up the sales function and staying in touch with the front line. What worked there works here too. Lean Practices can be applied successfully in any environment. This book will show you how!

Hong Chen, Vice President Sales
Business Region GCN, Festo

When I read through Liam's book, I couldn't help but smile how the writing mirrored the author—authentic, no-nonsense, straight to the bull's eye! As his assistant in China, I witnessed how Liam turned a gray, dull, and old-fashioned Braun plant into one of the most shining, vibrant, and modern factories even in the P&G world! With his strong determination to fight all bureaucracy, and relentless passion to drive efficiency, he has been a true leader who challenged the people around him, including myself, to stand up to challenges and step up. I believe this book by Liam (full of simple, common-sense but true solutions) can also help you get to the core roots of Lean and help your site to survive and thrive in this changing world!

Ana Xu, Human Resources Director, Asia, Voith Turbo

Liam Cassidy is a born leader. From this book, you will see how he led his factories towards the highest level of Lean Manufacturing in America, Europe, and Asia. He made all his manufacturing sites world class. I was fortunate to be hired by him to lead the EHS programs.

Liam knows how much an engaged workforce can play a role in efficiency maximisation. During his daily plant tour, he would engage with all employees he met from different levels and made them relax and enjoy and be happy at work.

In 2008, we held our first Family Day in the factory's history in recognition of our employees' great contribution in making our factory world class. Over 1200 employees attended the whole day activities together with their families. This made the whole factory one family and one team as a cohesive force.

Liam's support and mentoring helped me grow and has helped me build a successful career, leading several organisations. Whoever reads his book will gain the power, just like I did.

Chris Wang
GM of Shanghai Yunqiang International Logistics Co.
Vice President of Zhejiang Chaopei Logistics
President of Maanshan Kangfeide People Resource Co.

I had the privilege to read Liam's new book before most of the readers, and it brought me back to the days when I was his Finance Director working side by side with him.

Not only is this a book about Lean Manufacturing, but it's also a book of how a leader successfully leads transformations. In Chinese there is a saying that "Example is better than Precept," and Liam exemplifies that proverb. In this book, he uses his own experience to show you a picture of Lean expertise and factory management game changers.

Liam is a warm-hearted leader with a determined mind, who doesn't tolerate crap and gets to the core of the issues. Readers of this book now have the opportunity to learn his secrets.

Lily Yu, Managing Director, Trading International,
China

Foreword

When I was Communications Director at the Lean Enterprise Institute, one of my most important—and enjoyable—responsibilities was to find companies and managers who were doing something new or notable implementing Lean Management Principles. That's how I met Liam Cassidy, the author of the book you're holding.

I had written many case studies of Lean applications in metal-bashing industries, which is not surprising given Lean's Toyota roots. But I heard that Liam had used Lean Principles and Tools to save a highly automated toothbrush factory that included a pull system triggered by a packaging line. I had to get out to Iowa to see this.

Liam was the new manager of the largest toothbrush manufacturing plant in the world, located in Iowa City. He was sent there to close it. Production would go to cheaper factories in China or Mexico.

But Liam was a veteran Lean thinker and practitioner who had turned around factories in the UK and his native Ireland. He knew that the competitive power of rapidly and diligently applying Lean Principles, with the participation of local management and employees, could save almost any factory, including this one.

The Iowa factory had some advantages. Product quality was good. A dedicated workforce met deadlines and got products out the door reliably. Plus, the factory was smack in the middle of the company's biggest market, North America, giving it a supply chain advantage over China and Mexico.

But there were serious hurdles to making a leap to Lean Management. Based on visits to the shop floor, Cassidy estimated that, at any time, at least 30% of the equipment was down awaiting raw materials, maintenance, or work orders. The supply chain from suppliers through the factory was unstable. There was no regular training for employees, except when new equipment arrived. There was some automation in the Moulding Department, but the level of manual processes was high; higher than at factories in Asia and Mexico, where labor costs were much cheaper.

Most surprising was the complacency. Quarterly town-hall-style meetings focused on good news, not problems. The factory lacked a communications process that let people voice opinions or report problems.

Factory or Hindenburg?

In fact, the plan for shutting down the factory was two years old but had been seen only by the former Plant Manager and Human Resources Director. The leadership team and 750 hourly employees faced the industrial equivalent of having a window seat on the Hindenburg—but no one told them.

Cassidy began the transformation by meeting with the entire workforce to explain the dire competitive situation. First, he met privately to win over the union head by explaining what he would do and to share information about the closure plan. He didn't promise people that the corporate decision to shutter the factory would be changed. He explained that it had a shot at survival if everyone supported and participated in the Lean transformation. He promised to keep them informed through every step of the change.

Next, he trained teams of employees, managers, and staff in Lean Principles and Tools. As part of the instruction, trainers asked people for feedback on four questions:

- What is working?
- What is not working and what are the barriers to that?
- What should we retain and do more of?
- What do we need to do to become world class?

Responses were pooled and posted around the factory. Then, implementation teams were organised to simultaneously:

- Reorganise the warehouse with a pull system using Kanban signals to the factory floor
- Create a Kanban system for internal movement of work-in-progress and introduce Backflushing
- Create a central store for spare parts
- Adjust resources to ensure maximum support for production
- Implement the 5Ss methodology to improve workplace organisation
- Establish an inclusive communications process
- Develop Lean capability at suppliers, including a Kanban system to the factory's warehouse
- Use the single minute exchange of die methodology to improve changeover speed
- Practice autonomous and preventative maintenance
- Automate intensive manual operations.

In two years, the Lean Transformation had:

- Reduced direct headcount by 38% (mostly by attrition and separation packages)

- Improved schedule adherence from 92% to +99%
- Saved $16.1MM in reduced costs
- Implemented a pull system of internal and external Kanban for all regular materials
- Rooted a philosophy of Continuous Improvement in employees and managers
- Increased productivity by 34%, which had improved to 55% two years later when I visited

During my visit, Liam explained the factory's background and challenges. But my tour was conducted mostly by area managers and employees, who showed me around, explained the system, and patiently answered my questions, particularly about how the pull system worked. (By the way, you can read the case study "Toothbrush Plant Reverses Decay in Competitiveness" on the www.lean.org website) This was a transformation being done *with* people not *to* them.

Proven Pathway for Positive Change

The bulleted items above, along with the steps delineated in Chapter Five's section on "Transforming a Manufacturing Organisation," form the core of an approach that Liam used to save ten plants in high-cost and low-cost countries on three continents during a 40+ year career. As you'll learn from the six case studies ahead, the pathway is customised to the problems and priorities at each factory. Most became benchmark or model factories in their companies. Best of all, the factories sustained the gains, except for one that was closed after a corporate merger.

Although Liam writes about his method for successfully initiating and sustaining change, this isn't a technical book. There are plenty of excellent ones on the details of Lean

Management concepts and tools. This reads more like a business memoir with practical and motivating reflections from a successful business leader and lifelong learner. Liam gives you a proven Lean pathway for successful turnarounds, but he includes insights for dealing with myriad other business challenges that you face, such as improving toxic work cultures, managing change, surviving mergers, rousing dispirited workforces, and dealing with today's downpour of data. His key lifelong lessons from each experience are conveniently organised at the end of each chapter.

Liam pays as much attention to the social side of implementing change as he does to the technical side. In the pages ahead, he shares many valuable insights about how to communicate, inspire, and involve managers and employees. He has a way of connecting transformations with people's instinct for survival and their desire for meaningful work, all of which you can adopt. But he doesn't mince words about how to deal with truculent union officials or managers who stand in the way of change needed for the common good of saving a threatened factory.

A natural storyteller, Liam weaves in humorous or poignant anecdotes about the personalities and places he encountered in Europe, Asia, and the U.S. Although all the factories faced unique threats, there were common qualities too.

"Workforces have the same needs all over the world, regardless of where they reside," he writes in Chapter Ten on the reasons why Change Management works—and why it fails. "They want to be respected, and to be acknowledged by having a voice that is listened and responded to. They want opportunity for education and development, and to be fairly remunerated so that they can provide security for themselves and their family. The wonder is why so many organisations struggle to see that ..."

Make Your Factory Great & Keep It That Way is an informative, entertaining, and succinct book, perfect for reading on the plane ride to a factory you've been assigned to save.

Chet Marchwinski
Bethel, CT, USA
January 2022

Acknowledgements

I am indebted to Dustin Bilyk from the Author's Hand for his editing skills, encouragement, and patience as he guided me through this book all the way to publication.

To Chet Marchwinski for his advice and recommendations and for writing such a flattering Foreword. I first met Chet at Oral-B Iowa City in 2004 when he came to write a Case History for the Lean Enterprise Institute on the turnaround I had led there.

For David R. Norris, Packard Electric, from Warren, Ohio, who is now deceased. He was the first senior manager to give me the confidence so I could begin to realise my potential.

To Joe Ryan from JMR Consulting Ltd., who has provided me with outstanding advice and support since first meeting him at Oral-B Newbridge almost twenty years ago. His deep knowledge of organisations, culture, and change has been invaluable to me on my journey.

I am indebted to Paul McCarthy, Principal Consultant of McCarthy Insights, Shanghai and Belfast. Apart from his marketing and training skills, his knowledge of doing business in China, with his wide range of connections, helped me to avoid the many pitfalls that exist when attempting to successfully establish a business there.

To Frank McCourt of Packard Electric, and later Braun Oral-B, for his support and wisdom when I was a young senior manager. Recently promoted, I was struggling to see a way forward. His practical advice always helped me find a way.

I would like to thank Dennis (Sam) Herraghty for his terrific contribution to the packaging innovations at Oral-B. His patented design process for a most challenging project is still being used around the world, over 20 years later. His many innovative and cost-effective solutions helped avoid otherwise expensive investments.

For Keith Johnson in London, who, as a Lean Expert, has supported me on projects since 2000 on three continents. His no-nonsense application of the Lean Toolbox, and the results achieved, are second to none.

For Clem Smith in Dublin, who has also supported me on projects in the USA and Europe. His team building, mentoring skills, and keen eye for talent were invaluable in ensuring success.

In memory of my good friends and team members from Packard Electric, Mark McDonagh and Noel O'Leary, both sadly deceased. For a time, we showed what was possible.

To Father Dennis McGettigan, Raphoe, Co. Donegal. For his life-long friendship to me and my extended family. His innate goodness and faith in the human spirit is so inspiring in the most challenging of times.

Finally, for my wife, Anne, who has provided me with unstinting support throughout our marriage as well as all the turbulence that life throws up, and our three daughters, Lisa, Aoife, and Clare, who have brought both of us such joy.

- *Liam Cassidy*

Contents

Introduction

Manufacturing is the backbone of so many communities across the world. If a reasonably sized town has even one factory, you can bet that a large number of locals will be employed, and for every person employed in a factory there is an average of *four* people providing services directly or indirectly to it.

The manufacturing industry keeps communities alive and healthy; they thrive with the opportunities a local factory provides. Alternatively, when there is a lack of jobs, young people move away and these communities decline.

Apart from a spell in the army, I have spent my entire life in or around manufacturing operations.

Those unfamiliar have no idea the range of career opportunities there are in a large organisation that manufactures its products. They are almost endless: production, planning, shipping, quality, engineering, various technical trades, finance, information technology, supply chain management, marketing, design, new product development, and on it goes! I am one of so many who have benefited from these opportunities. Over the course of five decades, I developed a successful career in the manufacturing sector that is so far removed from the ambitions I had as a youth growing up in the hills of North West Donegal, Ireland.

However, I have also seen communities in my own country, as well as throughout Europe and the USA, become devastated by the loss of their factories. Often due to rising costs and a decline in competitiveness, these factories were transferred to lower cost locations. Sometimes there were supply issues, lead times that were too long, and, in many unionised environments, restrictive practices that made the operation financially unviable. Often, when those factories attempted to make necessary changes for the health of the site, the unions would resist change and the factories would shut down or move their operations elsewhere.

As my knowledge of how best to manufacture grew, I realised that communities losing their factories was almost always preventable. If factory leadership did the right things, their site would survive and thrive—I saw this over and over again with my intervention. It is true that events like a merger or a sudden unexpected market shift can impact a factory's well-being, but if the problems are costs, lead times, and/or an unmotivated workforce, all of this can be turned around and then maintained by enlightened leadership. I have seen so many factories close that I strongly believe could have survived with the right people in charge. If your factory is in trouble, understand this:

It *can* be saved.

Since becoming a Senior Operations Manager in the 1980's, and later shifting towards consultancy, **I have turned around *ten* manufacturing operations in Europe, USA, and Asia**, and I will teach you how to do the same. Moreover, if your factory is running well, you will find methods in this book that will *keep it that way.*

This book will take you through my working life, all the while providing you specific and actionable case studies where you will learn from both the successes and the failures. You will see how much has shifted in the manufacturing sector since the late 1960's, which will illustrate how practices in factory management across the world have changed and will always continue to change.

This experience has given me a unique perspective and has similarly allowed me to develop a unique approach to factory management, making it possible for me to transform poor performing plants on several continents in both high- and low-cost countries. These plants all became benchmarks for their wider organisations, and I saved several from closure.

If you are aware of common practices in the industry, then you will be familiar with the Lean Manufacturing philosophy. I first discovered Lean (then called World Class Manufacturing) in the mid-1980's. I have seen just about every approach there is to running a successful manufacturing operation, but I know of no better way than the Lean Philosophy and Toolbox. These I use as my foundation.

Some of you reading this may have already tried implementing Lean and failed. It is true that it has a high failure rate, but there are reasons for this. Without an unrelenting vision and strong—sometimes ruthless—leadership, the tools alone *will not* deliver the desired results. This you will see play out time and time again in my case studies. I will show you how to avoid these pitfalls and make the Lean Toolbox work for you.

If you've been working in factories for much of your life, then you know much is made of the Toyota Production System. It is credited with the Lean philosophy and thinking that has spawned hundreds if not thousands of books, videos, papers, and workshops. There is no doubt that Toyota is a great organisation. For me, they are the best manufacturer of cars in the world.

However, I have learned that it is not possible to transplant cultures such as Toyota's, and I have seen many attempts fail. Work life is so different in the West from that in Japan that I believe it futile to even try. In many cases, there is an overemphasis on the Toyota culture and slogans that confuse people and turn them off the actual program of Change. For me, if organisations I work with wish to use them, fine. But if not, I am quite happy to say, "Let's go see the workplace," as opposed to "Let's go to the Gemba."

I would even argue that much of what Toyota does and what makes them great is common sense. I will comment on that particular "sense" later, but it's important to note that some of their most successful tools they've been credited with inventing have been around for ages.

So many things can influence the local culture in organisations. I have seen marked differences in sites that are based in the same city, never mind another part of the world. I guess I have learned that even though so many of the same issues are occurring in organisations, their local culture, laced with their HQ's culture (wherever *that* resides), is influenced by local practices and thinking that has developed over years. It is also influenced by strong individuals who have worked there in the past, and others who are there presently. There is so much that goes into culture that simply transplanting a successful culture (i.e., Toyota) from one organisation or city to the next is near impossible.

The pathway to a successful Lean future for your organisation must incorporate all of this into the way forward. This is what I have learned to do, and this is what has made me successful. There is no "one way." Your pathway to success must incorporate Lean Thinking and its Toolbox, but you must also deal with the local issues that are unique to your site.

This book will not make you an expert in Lean Manufacturing, nor is it meant to. Specialised training is needed for their Toolbox. For many, a major mindset shift to its philosophy is required. Lean focusses on putting people at the very centre, investing in their ongoing development, and involving them in the design of their workplace. Using Lean, you will give your employees a voice at every level, then listen and respond to it.

In my experience, this is the most difficult hurdle to overcome. Many in leadership positions cannot make that shift to an employee-first attitude, and this is the primary reason for the large failure rate.

The best way to become familiar with the Lean Tools is to be part of a workforce that is using them. There is no better way. Of course, there are all colours of Lean belts offered (for a fee) online these days, which pretty much anyone with a mouse and keyboard can obtain. However, these are useless without extensive hands-on practice within a Lean workplace. I note that one of the most popular online courses combines Lean & 6 Sigma. I know what Lean is, I know what 6 Sigma is, but I do not know what the two are together! Lean is the complete package and does not need to be confused or combined with another system.

This book will show you how Lean, with the right kind of leadership and necessary ingredients, will truly transform your factory. Furthermore, with the right discipline, you will keep it that way.

If you are looking to read a book from another academic, this is not it. There are hundreds of books and articles published each year on how we should improve our organisations; many from academia. Some are worthy of study, but in my opinion, most of them are of limited practical use.

Something that has always puzzled me is how academics, with a handful of exceptions, can advise us on how to lead, manage, and behave, but they themselves come from organisations that are the opposite of what they preach. Many of their practices are archaic, and they will resist change like the most militant worker's unions.

There are also plenty of books from people who have walked the talk, and I have enjoyed many of them. However, I do believe that my experiences, including how I started my working life, learning as I progressed, and eventually discovering and embracing the Lean Philosophy before applying it successfully on sites all over the world, will be helpful to those readers struggling to transform their factories. I deliver a no-nonsense approach that you will find incredibly valuable.

It's been decades since I first stepped on a factory floor, but I still retain the passion that I had in the late 1960's. The thrill I get when I see organisations improve, and workforces, at every level, becoming enthused by a workplace atmosphere offering them far greater freedom, expression, and opportunity than found in traditional organisations, is as strong today as when I first became a senior manager. I hope to convey to the reader a sense of how on-site learning, often when you do not realise it, is more important than formal learning when shaping the person that we become.

I struggled for years in poorly supported, negative environments, sometimes not knowing if I would have a job next month. These kinds of experiences have made me who I am today. Every employee has the right to be supported, fulfilled, and given the opportunity to grow, and despite some tough times, I was eventually provided those opportunities. For those of you who are working in negative environments, I hope my story will provide some encouragement. For those of you who are implementing Change and endeavouring to improve your areas of responsibility, I am certain you will find direction within these pages.

CHAPTER ONE

A Sense of Place & Military Life

I was raised in a rural community in the hills of North West Donegal. We lived about a mile from a small village of around 200 people, surrounded by hardworking hill farmers who were great neighbours. Most people in my area were poor, but we were all sustained by music, humour, and a strong sense of community.

We were a military family, but as kids growing up, we worked the fields for our neighbours. Our mother bore eight of us at home with the help of a local district nurse who arrived on her bicycle soon after word had reached her.

Mother worked miracles raising us, with little in the way of amenities found in homes today. She was one of the "go-to" women in the neighbourhood. When babies were being born, harvests needed to be reaped, or bodies needed to be laid out in preparation for a funeral, Mother would be sent for. When a crisis arose in the neighbourhood, she would be there. To this day, I and my siblings still marvel at how she could carry such a load with such great humour and wisdom.

School was difficult and unrewarding for me. Many kids

suffered beatings in those days, and there was nothing any of us could do to stop it. Fear and resentment are my abiding memories, and I dropped out of the system when I was just sixteen years old. It left me with the impression that education was something to be endured rather than enjoyed. Only later would I learn that education is a joy when delivered in the right circumstances.

Reading was my salvation in those days. A neighbouring town had a small, two-roomed library. At one stage, I was taking four books per week and travelling the world in my imagination with the characters I encountered within. These were mostly about travel, adventure, and war. My grandfather was a professional soldier, served in India, and fought in the Boer War. He also served for a time in WWI but was discharged due to ill health. He died in 1922, so we only knew him from the stories that were passed down about places he had been and the adventures he had experienced. These helped fire my imagination as a young teenage boy.

So, with military life in my blood, when I left school at the age of sixteen, I too decided to become a professional soldier. It was my dream to join the army, travel, and see the world. Like most teenage boys, I had a taste for adventure, and so I lied about my age and enlisted on a three-year contract, joining the artillery regiment in the Irish Midlands.

The barracks was situated in Mullingar, which had a population of about 3000 back then, but has since grown to over 20,000. The barracks was built by the British in 1815 and could accommodate up to 1000 troops, if necessary. Rooms, or billets as they were called, housed eight of us per room—four beds on each side. Facilities were basic but adequate with each building containing their own washing and shower facilities. Hot water was provided just twice a week, so plenty of us were forced to endure cold showers in the middle of an Irish winter, particularly after a tough route march or a game of football on a muddy pitch.

Our canteen served breakfast at 8.00am, dinner at 12.30pm (quite common back then) and evening tea at 4.30pm. A typical breakfast would consist of eggs, bacon, sausage and tomato, tea, bread, and butter. Our dinners were often roast beef or chicken, potatoes, and vegetables. Evening tea was similar to breakfast. These days, healthier options are on offer and menus vary much more. Luckily, we had a separate canteen where we could purchase snacks and other essentials like toothpaste, soaps, and stationery.

Our initial three months basic training weeded out those who were not going to cut it, but the majority of us enjoyed the army and the variety it offered. It was tough and physical, but it taught us how to take care of ourselves and, in many cases, how to depend upon and support each other. I learned so much during those years that I would draw on later in life, and even though I didn't realise it at the time, it was here that I learned the value of teamwork, discipline, and self-reliance.

My first opportunity to travel abroad came just a year after enlisting. I was one of 350 men (no women served in those days) who travelled to Cyprus to serve with the United Nations Peacekeeping Force tasked with keeping the Turkish and Greek Cypriot warring factions apart. This was a terrific experience that involved travel, adventure and, when they occasionally opened up on each other, a whiff of danger.

Facilities at Cyprus were basic. Six of us lived in one tent, and we had just eighteen inches of space on either side of us. In this kind of environment, you quickly learn about discipline, tolerance, and patience, but it was also here that I learned one of the most valuable lessons that I carried with me for the rest of my life:

A place for everything and everything in its place.

Space was so tight that it took great discipline to keep your stuff together so you weren't encroaching on the guys next to you. We had a kit bag under our bed which contained our underwear, sports gear, towels, and personal bits and pieces. It needed to be in a certain spot and facing a particular direction, or we were reprimanded. Boots and shoes were placed alongside it. Behind our bed, our uniforms hung on hooks inserted into the canvas.

We spent two of every four weeks in mountain outposts observing troop movements and searching vehicles. Here, facilities were even more basic. We had a two-ring gas cooker for six men, but occasionally gas would not be delivered, and we had to resort to cooking over an open fire. This was challenging for those not used to it, but the hunger of a half-dozen young men was a great motivator.

We didn't have electricity, so we made do with oil lamps. These were insufficient for reading after dark, so readers would use a torch. Rations were delivered most days by land rover, but one of our outposts required a two-mile trek down a steep mountain to meet a man and his donkey, who would then bring our rations back up to our camp. There were occasions when rations weren't delivered at all.

We once went several days without rations due to an extended incident that started with a burst of gunfire occurring within the scope of our outpost. We were meant to radio into HQ which side was responsible, and they would use their contacts within those communities to put a stop to it. However, due to the echo effect of the valley, we were unsure who it was. So, in less than an hour, both sides were blazing away at each other, and the incident quickly spread to other parts of the island, keeping UN forces pinned down and curtailing many of our activities.

At such times, you learn to improvise. We spaced out what food we had and reused tea bags until they were nearly breaking apart and flavourless. We bought a kid goat from a local farmer, killed it, and cooked it. When times got tough, our sergeant would not tolerate anyone complaining. Anyone heard doing so would end up doing unpleasant tasks or fifty press-ups. These tasks included maintenance work on the trenches or cleaning out the latrines. Our sergeant had no shortage of ugly work to hit us with. If he didn't have anything for us to do, he would make one up!

I didn't like every man in our unit, nor was it expected that you did. There was one man in particular I disliked because he bullied younger soldiers like me. He was on the boxing team and used those skills inside the ring and out, but I never backed down, and we came to blows several times.

I found that, when you stand up for yourself, you gradually earn respect from your peers, and after a while this man left me alone. This happens within the workplace as well, often with quiet employees.

If you're one of these quiet employees or managers, don't let your workplace take you for granted. Voice your opinion. Don't be pushed around. I know this can be difficult for younger employees, but this is how respect is earned, even today. Quiet people who allow themselves to be taken for granted rarely get promoted, and this is true up and down the ranks of every organisation.

From time to time, a debate occurs on LinkedIn about teamwork. A strong thread of opinion argues that, for a team to be successful, they must like and respect each other. Respect is *essential*, but it must be earned by each member's contribution and behaviour. However, just like in the military, liking each other has nothing to do with success.

I have seen people on teams detest each other but get on with the job in hand. This was amplified for me in the army. Everyone had a role to play, regardless of how you felt about one another. If you failed, the overall objective of your team was in jeopardy—you were letting them all down because, for whatever reason, you had not delivered. In the army, letting your team down can lead to a loss of life. In the workplace, letting your team down can lead to a loss of one's livelihood.

Likewise, I played soccer for many years, and I have seen some of our players physically attack each other after a loss because someone let the side down. Occasional, honest mistakes were tolerated, but poor teamwork invoked a ferocious response.

Teamwork is essential for success within the workplace. Respect for each other is just as important, but it is *not* necessary to like everyone. Indeed, not doing so can often give the team dynamic an added edge.

I also learned that trying is better than not trying at all. In the military, they never gave you a choice, and this was never more the case than on our assignment in Cyprus.

At one point, our camp backed onto our own private beach. At first, we had no restrictions on swimming. That was until one of our guys almost drowned and had to be rescued by a local fishing boat. After that, swim hours were restricted to a schedule and there had to be a lifeguard on duty. We didn't have enough of those, so a number of us had to be trained.

My name appeared on the list of people to go through a week of intensive lifeguard training. When I told my platoon sergeant that I was barely able to stay afloat, he stared at me for a moment and said, "I didn't ask you if you *could* swim, but you will be by next Monday evening." He had erred in selecting me, but he was a proud man and did not wish to admit it. As for me, I was frightened of water, and even the other non-swimmers were streets ahead.

But he was right! By the end of the first day, I was swimming, and by the end of the week, I was quite proficient and became an official lifeguard to the camp. During the training rescues, it was a struggle to keep heads above water, but I surprised myself and others by my quick progress. Thankfully, I never had to make a real-life rescue. But I would not have been afraid to attempt it, and when I returned home from Cyprus, I was a confident, proficient swimmer.

That single lesson instilled in me something important that I have always carried with me. You're reading this book because of it!

Don't be afraid to have a go at challenges.

You may not be perfect, but you *will* improve. We are built for it. Besides, it's better than sitting on the side lines.

In total, I spent twelve months in Cyprus; two six-month periods about a year apart. This was a dream come true for me and remains one of the great experiences of my life. I also visited Jerusalem when it was still the capital of Jordan, and I spent a short time in Beirut. I got to see the Palestinian refugee camps there that later became hotbeds of recruitment for the Palestine Liberation Organisation and, later, Hamas. Even then, as young men witnessing the conditions the Palestinians were living under, we knew there was trouble ahead.

Often, with those that I have worked with, or when coaching young future leaders, I frequently refer to examples from those days when discussing leadership. In a military environment it is clear, *very* clear, on what your role is. You are trained in a task over and over until you are competent to do it at an acceptable standard, and you are then held accountable for it. Of course, military discipline cannot be directly applied to civilian life, but we should not lose sight of the fact that we still need a hierarchy. Employees need to be trained and supported, but then should be held accountable for what they deliver. From my experience working around the world, this is a huge issue for many organisations.

I completed several courses during my service, primarily in various types of armaments. The most notable was an eight-week intensive course that qualified me as an Artillery Surveyor. This brought me to technician level and a higher pay grade, and although it was of limited use in civilian street, the crash course in maths was useful when I moved on. I also left the military with my slide rule (the precursor to calculators) that was used in logarithms, roots, and trigonometry. Interestingly, it would also do multiplication and division but not addition or subtraction. When I moved on to manufacturing at Smiths Industries Ltd, calculators were gigantic and only available to senior managers and some accountants. So, my slide rule became the envy of many, and there was a request at Smiths that they be purchased for certain groups. This was declined due to the cost.

When my contract was up, I faced a choice: stay in the army or try something else. I was interviewed by a Barracks Commandant when I said I was considering leaving. I told him I would stay if he could guarantee that I would spend two of every three years overseas. He could not commit to that, so I left with a friend with the exciting plan to go to Australia.

But that never happened. Why?

Because so much of the action was in London.

Lessons learned in the Army that I carried with me for the rest of my life:

- ➤ The value of teamwork
- ➤ A Place for Everything & Everything in its Place
- ➤ Have a go; don't sit on the fence
- ➤ The value of discipline
- ➤ There should be consequences if you don't deliver what you are supposed to

CHAPTER TWO

Into the Manufacturing World: Smiths Industries

In this chapter, I will explain my introduction to manufacturing and what I learned during this time. I was not yet 20 years old and knew nothing about what I was getting into. However, what I learned early in my career on the floor was invaluable in providing me with the knowledge to progress my career. Moreover, the skills I learned in the army set me on a career path that was to take me all over the world with increasing success.

At Smiths, I learned about the role of the key support areas to production and how they must gel if the plant is to be successful. I learned that when the areas don't work together cohesively, the performance of the site will always suffer, and the stress levels of employees, at all levels, will increase dramatically. This was the case with my first employer in manufacturing.

As a young, single man in London, I had little idea about what I wanted to do with the rest of my life. All I was sure about was that I wanted to do something that interested me. As was common for many Irish people in my generation, London offered more opportunity than back home, both in terms of career choice and the amount of money you could earn. Moreover, London had a terrific social scene. I played soccer at a great club and made friends that I still have to this day. Later, I met my future wife there.

Smiths Industries Ltd were a large UK organisation with about ten factories in England and Wales, making speedometers and oil and pressure gauges for the auto industry. I was based at their factory on Cricklewood Broadway which had approximately 1200 employees.

The buildings were a mix of pre- and post-war, and they were maintained to high standards. Most importantly, as a young single person with little money, they had an excellent canteen that served breakfasts, lunches, and evening meals at very reasonable prices.

In those days, many organisations had large boards erected outside their buildings listing various jobs that were available. Walk-in interviews were commonplace, so my friend and I entered the building without much of a clue what we were applying for. The interview took no more than twenty minutes. We signed contracts right away and walked out wondering just what we had signed up for. Little did I know that this split-second decision would change the entire course of my life.

My first job at Smiths was to clean the materials used for the speedometers by dipping them into a specialised fluid. The process was carried out in a relatively small room where people would drop off the materials on designated shelving. When I arrived, the place was a mess. The storage areas hadn't been cleaned in ages, and I could only imagine what my platoon sergeant would have made of it.

I spent the first few days getting my area cleaned up and reorganising the storage shelving. I carried out a process I had learned in an army stores, which is similar to First in First Out, but this was before I had ever heard of FIFO. I also had a separate, simple process for urgent jobs, which made users much happier. I found an unused trolley with three levels of shelving, cleaned it up, and put an "Urgent Items" sign on it, then placed it close to my workplace. Previously, they had been frustrated with the erratic nature of the service. Now they knew what to expect.

With everything organised, I had loads of time on my hands. The Scottish guy I took over from had told me he couldn't handle the workload, but it was only because he was badly organised. People told me I was doing a great job, but in truth I felt awkward and a bit guilty because I felt like I was being lazy. I had done nothing special and had just applied some common sense.

It was then that I began to see that, for reasons that still challenge me to this day, common sense is not so common after all. People who are sensible outside of work and make good decisions about all aspects of their lives can behave the opposite way in the workplace. Since then, all over the world, I have encountered the most ridiculous situations that cry out for simple common-sense solutions that would immediately improve results. In this case, all it took was some simple organisation—something my platoon sergeant had hammered into me.

After a couple of weeks on the job, someone came into my room and introduced himself as a senior foreman. He said he had heard positive things about what I was doing and added that he was particularly impressed with the way I had cleaned up and laid out the workspace. He asked about my background and became very interested when I told him I had just left the Army. He left and returned minutes later with his boss, the Works Manager. He asked me some questions that were mostly centred around whether I would like to progress in the organisation. He asked, "Would you be prepared to go to technical college if we funded it?" Of course, I would!

Within a week, I was moved to the Management Training Department with about 30 other young people, and I commenced a program that included a range of manufacturing operations subjects. This included Work Study (later to become Industrial Engineering), Production Control, Production Management, Warehouse Management, and Quality Control. We had a short introductory course in all of them, then we would choose, with management's approval, what we wished to specialise in.

My choice fell into my lap. An urgent request came from Production to temporarily replace a supervisor who had taken ill. I was asked if I wanted to volunteer, and suddenly, after just a few months at Smiths, I found myself responsible for a production line with 35 people manning it. I was still a teenager!

In those days, the entire line was populated by women, while the off-line support and higher grades were dominated by men. Of course, this was unfair, but that was the way then, and it was largely unquestioned. Looking back, there were seven or eight strong women on that line who could have done a better job than me, but it would take some years before that began to change.

The Senior Foreman told me it was my job to hit the targets every day so that they earned their bonus and ensured that shipping targets were met. I could do that by meeting the schedule, make certain they had all the materials they needed, and by getting their problems fixed quickly. It sounded simple but it was anything but.

At this point, my official job title was Trainee Supervisor. Of course, up to this point in my career, I had no real comprehension of what was required of a manager, nor any real inkling of the goings on within the manufacturing world. My section head was supposed to provide support, but he was tied to another line due to staff shortages, so I was pretty much on my own.

It was to my great benefit that a couple of strong women on my line took me under their wing and were enormously helpful in those early days. They taught me how to handle difficult changeovers, gave me early warnings on shortages, and showed me how to quickly combine workstations when we were short of personnel so that the line could continue to function. This training from these women helped me immensely later in my career when I began developing Lean Workcells.

I also learned a lot from some of the older, wiser heads. There was one time I got a bit aggressive in the toolroom whilst trying to get someone to fix a problem. They made me wait, and I could see that they were enjoying it. Later, their foreman, who had fought in the desert with General Montgomery in WWII and was only a year or so from retirement, took me aside and gave me a lesson I have never forgotten. He said:

Regardless of how right you are in your demands, unless you build relationships with those who are doing the actual work, you will never get the full benefits of what they can provide. If you speak to them with respect, they will respond.

It was simple common sense that a young man like me needed to hear, and I have tried to practice that and teach it to others ever since. In time, I did go back and apologise to the guys in the toolroom, and we got along much better after that.

Although the introductory class training I received at

Smiths Industries was excellent, it was all high level and based on a lot of theory. It lacked follow through. There was very little in the way of training for Production Supervisors or management generally. I wasn't aware of them having any management development programs.

Moreover, many supervisors and managers had favourites within their teams, and this was a regular cause of contention. I too fell into that trap for a while. I heavily relied upon two women on the line, and they manipulated me into giving them overtime that wasn't really needed. Soon, they even began asking for extended breaks, using a variety of excuses. I struggled with it for a while but eventually said enough was enough and put a stop to it. I became conscious that it was impacting my relationships with others on the line, and they were waiting for me to do something about it.

Later, a couple of fellow workers said that the same two women tried taking advantage of every supervisor they worked for, and they wondered if I had the ability to handle it. I believe I got more respect from the crew after I dealt with the issue. Most surprising was the reaction from the two who were messing me about! They knew they were taking liberties with me, and when I put a stop to it, our working relationship improved instead of souring.

However, it took quite some time for me to learn to stand my ground, and ever since, I have been convinced that nobody should supervise or manage a team, no matter how small or large, without basic training on how to lead, motivate, listen, and deal with difficult situations. These are the basics, and if you don't have them, employee performance will suffer, and subordinates will lose respect for you.

At Smiths, I also learned a lesson about bonus systems that I have carried with me throughout my life. It is quite simple:

Avoid individual or department level systems!

Many issues impact on performance, and if they lead to reduced bonus levels you end up with all kinds of disputes. Keep it at site level and make it dependant on the *site* hitting their targets. Anything below that can cause havoc, as demonstrated by the following story.

One of my tasks as supervisor was to report the numbers produced each day and pass them to the materials planners as well as the person who tabulated all the numbers for payroll. This person in payroll was a small, fiery Welshman who could be very impatient—especially with younger people if he thought they were getting too big for their boots.

As I became more familiar with the work, I was finding it difficult to reconcile the inflated individual and department bonus numbers coming from the final workstation. I mentioned this to my section head a couple of times, but he just passed it off. Finally, I figured it out.

Each shift, approximately 10-15% of our rejects were diverted to a rework station, repaired, and then fed back in. This was normal, but what wasn't normal was that the rejects were being counted *twice*, which inflated the bonuses received by all employees on that line.

I checked the rules carefully and this was clearly not allowed. So, I went to the fiery Welshman, and he quickly and firmly told me to ignore it and get on with things.

But when I persisted, he told me to go to the foreman, who looked at me wearily for what seemed a long time and said, "Do you want to be the cause of a strike?"

I said, "Of course not!" He said if I continued on this vein, I most certainly would. He then went on to say that some systems are not perfect and must be worked around, but I wasn't able to officially acknowledge that.

This was my first small taste of the power of trade unions, but it would not be the last!

The problems seemed to increase the further I immersed myself. There were other issues that were similarly ignored, including stoppages for technical support, unplanned changeovers, and much more, even though our procedure said otherwise.

I was never comfortable with that. I spoke to other supervisors about it, and they just said that's the way it was. I felt there was something wrong when both management and the general workforce had an unspoken agreement to ignore certain things, and it provided me with a sharp lesson in what to avoid if ever establishing bonus systems.

Another aspect of the environment within Smiths Industries was the rigid hierarchy. In many ways it was more pronounced than in the Army. The entire system was designed to make it clear, *very* clear, where everyone's place was on the ladder.

In Operations, there were long and short, white and brown coats which distinguished the various levels of trades and management. For senior Operations personnel there was a short pinkish one. Senior foremen were issued an underarm zip briefcase, and the next levels up were provided a brown leather case. Amusingly, with every promotion, many would upgrade the newspaper they bought! *The Daily Mirror* was the shop floor paper of choice, then as people were promoted, they graduated to either the *Daily Express* or *Mail*. For senior managers it was either the *Telegraph* or *Times*.

There were toilets for different levels too, and it was seriously frowned upon if you strayed into the wrong one. Likewise, there were different arrangements in the canteen. Workers would line up for their food, while more senior ranks would sit in a roped-off area and be waited upon. Moreover, it would be unthinkable to enter a senior manager's office without an appointment. They had to be addressed as "Mister," unless, of course, they were female. However, I can recall only one female senior manager and she was in Human Resources, then called Personnel. Such were the times.

I found the classism to be petty and childish. Likewise, many young people were beginning to challenge and question why it had to be like this. The culture was shifting.

I should acknowledge, however, that in all my time in England, I was treated with the utmost fairness, and indeed upgraded and promoted several times. The fact that this occurred during the IRA campaign of bombings and shootings in cities and military bases throughout Great Britain and Europe is a testament to the fairness and integrity of employers there.

I was with Smiths Industries for two and a half years but began to get restless and wondered what else I could try. Jobs were plentiful in those days and opportunities were everywhere.

Given the materials supply mess I was dealing with daily, I began to think about the supply chain. I had been having conversations with people in Smiths' Materials Department, and they were full of ideas on how to improve things but frustrated that nobody in management would listen to them. While I never understood why that was the case then, it's something I have witnessed for much of my working life. I cannot say that it has got much better.

So, when I saw a job advertisement for Planners at AC Delco (a division of General Motors), I applied for the job immediately. Thus began my next foray into the manufacturing sector, and one that would test me for many years to come.

Lessons learned at Smiths Industries that I carried with me for the rest of my life:

- ➢ Overemphasis on rank is unhealthy and stifles employee engagement and talent
- ➢ Avoid department and individual bonus systems
- ➢ Do not have unspoken agreements with unions or employees whereby certain issues are ignored
- ➢ Treat everyone fairly; do not have favourites
- ➢ Ensure that anyone who leads a team first receives basic training
- ➢ A chaotic Materials Department will consume resources at every level, damage relationships with customers, and add a lot of non-value-added activity

CHAPTER THREE

Case Study: AC Delco
A Division of General Motors

Once again, my military background impressed at the AC Delco interviews and, along with my recent production management experience, I skipped the trainee and junior steps and jumped straight to Senior Planner.

I was responsible for about 2500 items. Fuel pumps, spark plugs, and filters made up the bulk of my responsibilities. My job was to produce a monthly schedule for several other GM factories who would produce and deliver to our distribution warehouse, where we would then ship to a network of distributors across the UK and parts of Europe. Later, we would supply a couple of the large auto organisations like Rover and Jaguar.

Our offices were in close proximity to the warehouse, so we had easy interaction with staff there. Dorothy, a middle-aged woman, was my assistant and had been doing that job for *years*, which serves as yet another example of an environment in which women were considered second-class citizens within the workplace. She should have been doing my job. She was competent, smart, and she was a terrific teacher. However, if she was in any way bitter, she or her colleagues never showed it. They left an indelible impression on me, and I've tried to pay that kindness forward ever since.

My desk was next to Tom, a helpful middle-aged man who was originally from London's East End and had the wit that area is famous for. He had fought in WWII, but he was the only one of ten ex-military in our office who talked openly about it. The rest never mentioned it and attempts to get them to do so would always fail.

Looking back, they were serious, quiet men who just got on with work and rarely participated in the usual office banter that took place. As a young man, I didn't see it then, but they had a certain nobility to them.

Dorothy maintained a Kardex system, where each item had a card dedicated to it and all receipts and usages were manually posted. It was simple and efficient for its time and was enhanced by colour codes showing fast, medium, and slow-moving items. It also included min/max quantities and shortages. The cards were stored in trays contained within a desk-high cabinet, placed alongside our desks. Each product group had its own section, and each item had its own individual part number stored numerically for easy access. For updating, we simply removed the relevant tray. It was brilliant in its simplicity with so much critical information within our control. This is in contrast to many modern systems where we often feel overwhelmed with too much information and a lack of trust in its accuracy.

I often comment that modern systems, with all their advantages, have stifled people's thinking when introducing materials management systems to their workplace. I have seen organisations with no more than a few hundred items introduce an expensive ERP (Enterprise Resource Planning) system that they just don't need. So many organisations have forgotten how to keep it simple!

It was around this time that computers were first being introduced, and each month we received a printout of our part numbers with a projection going out over a rolling 12 months. We quickly learned to ignore it and applied our local knowledge and common sense when putting our demands and forecast together.

Today, despite all the progress with sophisticated systems, the situation remains largely the same in many manufacturing operations. Forecasts remain unreliable, reflecting the overly optimistic viewpoint of sales and marketing, but are kept realistic by experienced planners. Many organisations do not pay sufficient attention to this. Planners need to have experience with their portfolio and must balance optimism with realism.

Dorothy was invaluable in that respect as she had an encyclopedic knowledge of our parts and knew the ability of our suppliers to deliver them. She was able to coach me to anticipate when extra shifts were needed at the factories, which was incredibly important, because adding shifts was expensive and wasteful unless they were truly needed.

It was here that, for the first time, I learned about smoothening the schedule when you saw increases coming. Instead of hitting the factory with a sudden higher demand, you build increases in gradually over a longer period of time. Despite all the advances in processes and systems since then, I still encounter this inventory management issue across the globe.

Later, in the Lean world, this would become known as Heijunka—a Japanese word that credits Toyota with its development. However, as I have mentioned previously, this method and others were around long before Toyota became famous in the West for their production system.

One of the more interesting aspects of my job was working with Marketing on sales promotions. Usually, when a campaign was being planned, the respective planner would be part of the team so that the additional requirements could be built into our forecasts. This worked really well, and it was before email, so we all had to be in the same building and meet face to face to work things out. Email is a wonderful, essential tool in today's workplace, but too often it becomes an unnecessary substitute for face-to-face discussions.

I recall one marketing campaign where I learned a couple of valuable lessons. One of the incentives that the sales team passed onto their distributors was a 5lb tin of ham. Everyone thought this was a wonderful idea.

I ordered 1500 of those tins, and after a couple of weeks, over 1000 still remained in the warehouse. The warehouse supervisor was soon chasing me around looking for a solution, because the tins were consuming a lot of his space. But I didn't have one. The marketing team had moved on and were already planning their next campaign. I was on my own.

Eventually, I got permission to offer them to employees, and while this shifted a few hundred tins, people didn't want more than two or three of them.

The solution was forced upon us. We went on Easter holiday. The weather was unseasonably hot over the break, and the cans began to leak and smell terribly. When we returned, the local neighbourhood was in an uproar. Employees were threatening not to come to work, and even the police were involved. GM did not appreciate the bad publicity and local media were sniffing around, literally.

My department manager told me to fix it and *to do so today!* So, I scrambled for a solution.

Compared to today, environmental controls were loose to say the least, and while I tried calling a couple of waste disposal companies, they needed several days before they could deal with it, which wasn't good enough for my boss (or the neighbourhood). Instead, I called a friend who owned a construction company that had trucks ferrying materials to a motorway being built. He carted the ham away for a nominal fee and pitched them into the foundations of the motorway. Of course, that would be illegal today, but back then it was quite acceptable. My boss was delighted and gave me a lunch voucher!

From that day on, any time I was involved in product promotion, I always clarified whether there was a sell-by date. I also learned to be wary of marketing optimism. It's their job to be optimistic, but planners need to be practical, so from then on, I ensured that promo items would be scheduled over a period, with a proviso that we could cancel some if necessary.

During my time as a Senior Planner at AC Delco, I felt like things were changing, and mostly for the better. But rigid hierarchical systems, where power took precedence over what was right, existed here just like at Smiths. So much time was wasted waiting for permission for the simplest of tasks.

For example, to get a single item from the stationery store, one needed to have a requisition in triplicate signed by a section head or a manager. Surprisingly, even today approval processes are one of the most frustrating issues within organisations—the most common being purchase approvals. The exercise of power, as opposed to common sense, still exists.

Today, when I work with organisations, I work with leadership to create simple processes that grant approval almost immediately. These are carefully designed to be appropriate for the various levels within the organisation, which don't only protect the organisation from unnecessary spending but it removes significant delay and frustration.

I believe that the best system of management still needs a hierarchy, but with the freedom of cross-functional activity and well-defined scopes of responsibility at every level. Approval should only be required when something needs to happen that exceeds that scope. When organisations do that well, it is like taking the handbrake off a car. Suddenly things are moving faster, and people respond positively to that kind of freedom.

It was around this time my wife and I got married, and we had to decide where to plant our roots. She is also Irish and was becoming increasingly anxious to move back to Ireland. Initially, I was hesitant about it. I was quite settled in London. Football was a big part of my life then, and I was playing with a great club. It also helped that work was going well.

Suddenly, however, GM announced that they were building a new plant in Dublin through their Packard Electric Division. My divisional director asked me if I was interested in moving back and, knowing my wife would love the move, I said I was.

Despite the growth I had experienced in London, I was excited about moving back to Ireland and bringing the invaluable lessons learned with me. I knew I would be confident when applying them to challenging and difficult environments.

However, it was in Ireland that I would learn about the negative effects of selfish and self-absorbed leadership, as well as the destructive power of trade unions. I would also learn, for the first time, how truly successful the Lean Philosophy and Toolbox could be.

Lessons learned at AC Delco London that I carried with me for the rest of my life:

➢ Hierarchy without regular cross-functional activity and employee involvement is just the exercise of rank and power

➢ Keeping women in admin and non-management roles is not only unfair and discriminatory but prevents the organisation from fully utilising its talent

➢ We don't always need expensive systems for effective materials control

➢ Beware of marketing optimism and use local planning knowledge to balance forecasts

CHAPTER FOUR

Case Study: Packard Electric A Division of General Motors

The Packard Electric factory was brand new and built on the outskirts of Dublin, nestled within an industrial area where several factories were already located. Directly opposite of Packard was a cigarette factory, and just up the road the famous Jacob's biscuits were produced. Surrounding the industrial zone were green fields with cattle and sheep, and their sounds mingled with the aroma of biscuits and cigarettes.

Over the years, Dublin would grow outwards and around the Packard Electric factory, and the countryside and local village of Tallaght would be absorbed into the fastest-growing city in Ireland. My wife and I found a small apartment in a suburb of Dublin about a fifteen-minute drive from the site. A year later, we would purchase our first home in a small town thirty minutes away.

Moving to the new plant was a pleasant change from the pre-war sites in London. I was now working in a brand-new, modern building that still smelled of fresh paint. Unemployment was high and good jobs were hard to come by in the area, so with this exciting new development, goodwill existed at all levels in the factory.

I arrived in the early days of the Packard Electric plant's expansion. At first, we had just 50 employees—a single shift and three assembly lines. However, over the following four years, this grew to thirteen assembly lines working 24 hours a day, and we numbered around 700 employees. Eventually we would peak at around 1300.

I was immediately struck by the lack of formality in Ireland; a very welcome change from what I had experienced in England. The plant manager and his team were friendly, and everyone was on a first-name basis. No appointments were needed if someone wanted to see the Department Head. You just went to their office.

Our main function was to build wiring harnesses for Opel and Vauxhall—also divisions of GM. Later, we would supply Austin Rover and Ford. Opel had sites in Rüsselsheim, Düsseldorf, and Antwerp, while Vauxhall was in Liverpool and Luton.

My first job was running the warehouse and shipping areas. Most of it was new to me, but I soon settled in. I had a team of six employees—which later grew to twelve—spread across two shifts. They were anxious to get on with things and needed minimum direction from me. A no-nonsense chargehand coordinated their tasks, carried out whatever training was required, and generally ensured we had a clean and well-run warehouse. Most of my time was spent on paperwork and liaising with shipping companies and customs to ensure clearance for both inward and outbound shipments. These were the days when just six countries were in the European Union, and the movement of goods across Europe required extensive documentation that was very different depending on which borders were being crossed.

I have worked all over the world and have yet to find shipping companies that are better than what we have in Ireland. The companies I dealt with over the years were just brilliant and nothing was too much trouble. Communications did not stop on Friday afternoons but continued over the weekends when necessary. I have often wondered why they are so good, and I guess it is because we are an island off an island and just had to get good at it!

I was in charge of obtaining customs clearance for inbound materials, and, except for my brief run-in with them at Smiths, this was where I first encountered the negative role of trade unions. Often, clearance of shipments was controlled at the whim of disgruntled local employees at the docks or airports, who were always supported by their union. They seemed to be in permanent dispute with management, which meant the dreaded "working to rule," which really meant anything they wanted it to be.

Strictly speaking, it is meant to be a very precise interpretation of job roles, where overtime would be prohibited. A typical response to many requests would be: "It's not my job to answer the phone," or "It's not my job to move that box."

Thankfully, unions have almost disappeared from the private sector in Ireland today. This collapse occurred because most of those businesses fled the country due to union behaviour or went out of business altogether.

However, over the past three decades, there have been many new entrants into the Irish manufacturing sector, and most are union free. Companies have kept it that way by practicing enlightened management behaviour, while having strong communications processes and all kinds of employment involvement models. They also provide conditions that are rarely available where unions exist.

A union's *modus operandi* is to place themselves between management and the workforce to prevent them from talking to each other. This is a model for disaster, and sadly this is what has happened to our public services here in Ireland and in many other places across the world. The unions are in full control of every aspect of our public services and determine what happens and what does not. Every change is paid for dearly. Unfortunately, much of what is negotiated and paid for is never implemented, and so the circus goes on and on. I call it the "Triumph of our Public Sector Trade Unions." Historically they've had massive control over what happens without any accountability.

In the 1970's, often the only way to get shipments cleared on time was by the processing of a brown envelope. There were unwritten rules about how this was managed. Typically, it was a "contribution" to a Christmas event or a retirement party for a long-serving employee. There were no shortages of "worthy" causes! This was blackmail, but of course nobody called it that. Unions controlled most of what happened at the ports and airports in those days, and such transactions were common.

Elements of our media and some of our well-known folk singers and poets fondly reminisce about those days and the characters that dominated the unions. Myself and many others certainly do not remember it with any fondness, and we did not mourn the passing of their dominance. We witnessed the negative use of power by trade unions at its worst, and this behaviour inflicted great harm on business and industry in our towns and cities all over Ireland. On the other hand, we hear a lot about the negative role of owners and management (much of it justified), but not enough about the role of unions in the demise of the industrial sectors across the Western world.

The Ireland plant was unionised from day one, and in the early days, employee and management relations were positive.

Everyone was new, jobs were scarce, and people were excited about this new venture.

However, a combination of weak and selfish management, as well as some shop stewards anxious to impress elements of the workforce, led to a swift deterioration. Macho shop stewards started to run with grievances. Some of these were real, but many were ridiculous. However, this appealed to some employees, and you only need a small number of disgruntled workers to create a lot of unrest.

At the time, this macho culture within the auto industry was prevalent at all levels. Supervisors were encouraged to simply tell employees what to do. Many managers operated similarly. This culture was reinforced by a production system based on the stopwatch approach that was popular at the time, where every task had a standard determined by an industrial engineer with a stopwatch and a clipboard.

They would observe the person performing the task and come up with a standard time. Often, they too had targets to reach and were rewarded by reducing the standard. However, data and formula were not shared with the employee doing the work, and this led to suspicion and resentment.

Industrial Engineering, as it was known then, had far too much power, and I believe this approach contributed to much of the unrest in employees not just in Ireland, but also the UK, much of Europe, and the USA. Sadly, these factors continued to be a feature for many years and were major contributors in the decisions to close sites all over the world.

Thankfully, this approach became outdated with the advent of new thinking called World Class Manufacturing (WCM), later renamed Lean. Basically, it is a structured, integrated system that includes all aspects of manufacturing. One of the key differences from the stopwatch approach is that it encourages the involvement of the workforce at every level. As I mentioned earlier, it was derived from the Toyota Production System, which is widely acknowledged to be the best producer of automobiles in the world.

However, in those days before Lean, a feature in every factory was the large amount of inventory that required storage. This added enormously to costs of all kinds, including storage, movement, counting, and space. There were many factories that failed—and many that still do—because of this alone. Later, I learned solutions that could be applied to this enormous problem with great success.

In the Packard plant, one half of the shop floor was occupied by Lead Prep. This was where cables were cut and terminals were applied in various lengths and colours, then fed to the other half of the floor to our assembly lines, called Final Assembly. However, the Lead Prep area was struggling to meet the increasing demands, and lost time became a regular factor in Final Assembly.

Moreover, tensions were emerging at senior management levels. My function head, Jim, was on assignment from the USA. He was primarily responsible for supplying our customers, and he became increasingly critical of the way Production was being managed. So, the Plant Manager, also from the USA, decided to split the responsibility for the shop floor in two and gave Jim additional responsibility for the Lead Prep area. Shortly after, the Head of Production was fired, and the HR Leader was transferred to take over Final Assembly. Jim asked if I would move into his new production area and take over the day shift, which was where most of the activity took place.

Initially, I was disappointed. The Materials Management/ Supply Chain areas were still my primary interest. The warehouse and shipping areas covered major parts of that, and I was still learning. However, Jim had been impressed with what I had done in my short time there and was anxious for me to go to Production. So, I returned to what I had being doing at Smiths a few years previously, and re-entered Production supervision with responsibility for Lead Prep.

A poorly organised staff and lousy scheduling were my first impressions. We had about 12 cutting and 30 terminating machines, all of which were being scheduled individually. Waiting time was endemic, and operators were wandering around and disappearing for long periods. At Smiths, I learned that where supply system processes were poor, you had to improvise as much as possible. You could stand back and blame the processes and systems, but that would not lead to early improvements.

So, I worked closely with the local schedulers, and we established a rough visual system within our Work in Progress (WIP) areas to trigger replenishment. This was before I had ever heard of Toyota or Just in Time, but what we came up with was just common sense.

I often tell young people new to manufacturing to think about how our mothers organised our kitchens. It is often the smallest room in the house, but it contains hundreds of items. By necessity, they had to practice a visual process, because they didn't have the space to play with, and they had to avoid food going bad. This is essentially Kanban.

The same process applies in the local corner shop or in your supermarket. Just watch the bread being delivered. The driver comes in, scans the shelves, removes yesterday's leftovers, and replaces the loaves with fresh ones. So simple! Toyota didn't invent everything!

Within a week or so, supply to our machines was stabilising, and in turn, so was our supply to Final Assembly. I also put in some sensible controls at each production point, where operators had targets and were filling in their production against those. I would follow up diligently when the targets weren't being produced, and suddenly, the shop floor began to perform better. After a couple of months, downtime was almost a thing of the past.

Another feature of this change was increasing levels of WIP. We were running out of space in our storage areas. For today's Lean practitioners this is hard to understand, but in the environment of the 1970-1990's, large inventories were normal. So, despite our storage space issues, Jim was delighted with this and encouraged me to keep building! He awarded me with a special salary increase and recommended me for a promotion.

In those days, the critical measurement of production was standard hours. In other words, you were supposed to maximise and keep every machine producing regardless of whether the inventory was needed or not. To shut machines down wasn't even a point to be considered.

Of course, this was crazy, because it drove demand all the way back to our suppliers, which meant they were sending us stock and overcrowding our warehouse with materials we didn't need. Moreover, we were consuming materials we *did* need on *non-urgent* parts, often not leaving any for urgent items. If this sounds crazy, yes it was. But several decades later, it is amazing how many factories around the world are still behaving like this.

In the meantime, the factory was growing at a hectic pace and recruitment increased rapidly. We were bringing in 20-30 employees per week without sufficient screening and recruiting supervisors internally with little or no training. The unions were beginning to exert themselves, and in the absence of a strong united management philosophy, industrial relations began to deteriorate.

After about eighteen successful months in Production, I was asked to transfer to Quality Control. We had a huge Quality

Department with three inspectors at the end of every production line, and my job was to integrate what they did into Production and have roving auditors instead. This was a positive move and in line with what modern organisations were doing. Before I joined the Quality team, these inspectors were correcting and highlighting errors made earlier in the process, which was hugely expensive. My role was to work with Production to have those root causes corrected so they could be removed.

My work in Quality Control was a nine-month assignment. It was quite sensitive, as the auditors, who carried a higher pay grade, were chosen from the inspectors, and only a small number were selected. The remainder would be integrated into Production which, to many, was unattractive compared to working in the Quality Department.

I managed to accomplish it without any major issues. The skills I had learned at Smiths and AC Delco, about building relationships and involving people in discussions, were instrumental in that success. Auditors must have a thorough knowledge of the technical aspects of their job and be good communicators. The best auditors work with Production to prevent problems before they happen, and they pay more attention to new employees and help them until they are comfortable in their role.

During this process, I also learned that, at every level, there are individuals who are influential with their peers. Once you know who they are, you pay them more attention. Convince them of your mission and you increase your chance of success with everyone.

At this point, I had three assignments under my belt with Packard Electric, all of which I was deemed to have performed well. With this success, I was offered a job in middle management, responsible for the Lead Prep area—this time for all three shifts. This was an easy assignment for me, and I found myself being involved in other projects, such as dealing with new products and obsolescence.

I was also continuing my professional development by attending courses at various institutions. In one of these courses, I first learned about Lean. This not only grabbed my interest, but it also completely changed my outlook on what manufacturing was.

I devoured any books I could get my hands on, and I visited any worksite I could that was experimenting with Lean. I was completely taken by the *process* approach, where we concentrated on how work gets done, rather than on the individual. In other words, everyone had a role to play in the larger operation, and the people doing the actual work were involved in designing how it should be done. It was the beginning of the end for the stopwatch approach and that heinous brand of Industrial Engineering.

There are many ways in which to describe Lean, but one of my favourites is by the Lean Enterprise Institute (LEI):

"The core idea is to maximise Customer Value while minimising waste. Simply, Lean means creating more value for customers with fewer resources."

I would add that it also frees up workforces from rigid hierarchies, allows everyone to have a voice and to be responded to, and then allows all employees to be involved in day-to-day decision-making and problem-solving in their areas of work. This was something we lacked at Packard for the majority of my tenure, despite the best efforts of myself and others to have Lean embedded.

After six months in Lead Prep, I was asked to go to Final Assembly. For a long time, Final Assembly had blamed their poor performance on supply from Production, but I had helped sort that out. When the excuses ran out, the incumbent left the organisation and I took over.

When I arrived, there were some major issues with employee production and motivation. But within three months, productivity was at an all-time high, and I had calmed the industrial relations environment. At that time, we had an enlightened HR team. We established a communications process with the shop steward team, and we had a couple years of excellent performance. So, even though the overall environment was not good, because we had senior managers more concerned about their personal ambitions than the future of the plant, it was still possible, at a local level with the right approach, to have a strong performing department.

However, I learned that it is not possible to maintain that long term with poor management. Too many negative factors outside of your control will eventually impact it.

The HR leader, Sean, was promoted to Plant Manager, and in turn he promoted me to Head of Production. I was 32, and quite young for this position as far as it went at GM. I joined a senior leadership team that was riven with division; some with personal ambitions far beyond their ability who were opposed to the new Plant Manager and me, as I was considered his up-and-coming protégé. Some broader perspective is required here for readers to get an understanding of the environment I am attempting to explain.

In Packard Electric's first eight years, we had five plant managers and several changes of department heads. Apart from a couple of notable exceptions, we were cursed with Human Resource managers who were overly focused on their careers first, while the needs of the workforce came second. Moreover, in my sixteen years there, we never had a developed management philosophy or a united leadership team. Little wonder the workforce paid so much attention to their shop stewards.

We were also unfortunate in our technical support departments. Again, a mix of incompetence and personal ambition took precedence over providing adequate services. Requesting improvements were met with suspicion and denial, and suggestions were dismissed out of hand. Operating in a challenging environment like this gave me the experience to enthusiastically embrace the Lean philosophy when the opportunity came.

Our plant manager, Sean, was undermined from day one by some of his leadership team, and when the industrial relations environment deteriorated, he came under increasing scrutiny from his German bosses. There were several visits from HQ when Sean was presenting results, and a couple of his team members visibly undermined him with their body language and whispered comments. Eventually, Sean resigned and left the organisation. He was replaced by Kurt, a German national who had been running a factory in Spain.

Sean had been a decent, sincere man with plenty of ability, but he was too trusting and forgiving. If he'd had a strong, united leadership team, I am confident he would have been successful, but too many forces were reigned against him. As he acknowledged in his farewell speech, he had not been ruthless enough and should have fired some of his leadership team.

It was a lesson I have never forgotten, and it's one that I've passed on to any would-be leader or team. No matter how well intentioned you are, even if you are in possession of all the talents and resources necessary for success, without a united team focussing on the same vision and working relentlessly towards it, you will not be successful. Moreover, you cannot carry anyone at a senior level, because if you do, either for lack of ability or loyalty, they will undermine your mission every day. Make all reasonable efforts to turn them around, but if that fails, separate them from the organisation for the greater good. Do not delay with this. A leader's first priority is the *greater good*. In other words:

Your priority should always be the overall success of the site, its workforce, and its long-term survival.

With Sean's departure, the industrial relations environment was getting steadily worse, and within a year or so of his arrival, Kurt came up with a different strategy. He replaced the Head of HR and decided to negotiate a new Labour Agreement.

Negotiations commenced, but by this stage some of the shop stewards had become drunk on their power. No matter how frivolous a grievance, and there were so many, it would be supported all the way though the grievance procedure, appeal after appeal, involving more and more resources. Hundreds of thousands of man hours were spent on debating trivial non-grievances, but they were indulged all the way to the final stages of the Federated Union of Employers and the Labour Court. This devoured factory resources at every level— resources that should have been focused on making the site more productive and attractive for future investment.

This behaviour was fuelled by the lack of a progressive employee relations environment and the selfish and self-serving conduct of too many senior and middle managers. However, the unions should not be let off the hook. Much of their behaviour was just downright destructive and led by a tiny number of shop stewards with chips on their shoulders, intent on disruption.

This was a very difficult time for me, as I had problems gaining the trust of the Plant Manager. I was running Production, so ostensibly I was responsible for the behaviour of the workforce there. I had the support of a couple of people on the senior team and a small but strong support team of middle managers, and while I had been delivering strong results and had confidence in my own ability, I couldn't continue to deliver results when repeated appointments at senior levels failed to provide the support Production needed.

I learned a lesson I never have forgotten:

The vast majority of issues on the factory floor originate away from it. You must have all departments performing well and working toward the same goals to achieve a site's potential.

Against this background, Labour Agreement negotiations were never going to go anywhere, so a strategy to prepare for a lengthy strike was quietly put in place. I had previously expressed to Kurt a desire to work in Materials Management. So, he called me one day to tell me I was transferring there but told me not to announce it yet, as he still had to deal with the current Materials Management Manager. He was a German on assignment and Kurt did not trust him to keep things confidential.

In the event of negotiations failing, and the strike going ahead, I was told to find a way to quietly have raw materials routed to a site about an hour from London, and another at HQ, near Düsseldorf, where we had set up emergency production centres to produce our final products. A technical team was put in place to prepare the centres. Meanwhile, our plant was told to produce ten weeks' inventory of product before the strike occurred.

So, this was quite a dilemma for me. I was not yet the Materials Manager, but somehow, I had to find a way to work with suppliers to route materials to two locations in two different countries and keep it a secret! There were *seven* suppliers involved.

I had become quite friendly with the new materials controller, Chris, who had previously taken over from me in the warehouse when I moved to Production. We both went straight to the military after school, though he had gone to the Royal Marines, which is widely considered to be one of the toughest regiments of soldiers in the world. After six years, he had resigned on principle after the military's involvement in what became known as Bloody Sunday—an incident in Derry when twenty-six innocent civilians were shot, thirteen of them fatally. He had not served in Northern Ireland but felt strongly he could no longer stay in an army that would allow that to happen and then defend it.

I felt I could trust Chris, so I confided in him and asked if he would be my key to routing supplies with no-one the wiser. He agreed.

For two months prior to the strike, he quietly set about successfully establishing secret supply routes. An alternative method of invoicing was even set up. He did a terrific job, and although he enjoyed success in the department for years after, he did not get the acknowledgment that he deserved at the time. I share responsibility for that. In general, our environment was not good at acknowledging great work. But years later, I took great pleasure in placing him at a couple of my factories in critical planning roles, where he helped bring some much-needed experience and stability.

As many of us expected, the strike went ahead in the harsh winter of 1987. We had the largest snowfall in over two decades and unusually cold weather for several weeks. It caused quite a stir in the media as the unions, trying to outflank us, walked out two days early, leaving a warehouse full of finished product. We had planned to maximise production up to Friday and then ship everything that evening. Instead, snow was heavy on the ground and pickets were put in place 24 hours a day.

So, the following week, we arranged for a convoy of trucks to come in the middle of the night when just a couple of people were on picket duty. We had the trucks quickly loaded and taken to a secret location before being ferried to England and then Germany. This was planned like a military operation, and it was executed perfectly.

In the meantime, the remaining staff, such as supervisors and office personnel, continued to build critical products. When a reasonable quantity of these were built, our plant manager, who had an inclination for being flamboyant, decided to bring in a helicopter to ship them out. So, one mid-morning, the helicopter arrived, and a team of us, including Kurt, loaded the chopper and away it went. Kurt even had someone record the scene with a camera. That grabbed headlines on just about every media organ that existed and caused a lot of debate about unnecessarily antagonising unions.

The strike lasted seven weeks before they finally went to vote on the proposed new agreement—one that had previously been vehemently opposed by the unions. It was accepted by a narrow margin, so everyone returned to work.

At this stage, a major opportunity was missed. When employees returned, managers and supervisors were told to "walk on eggshells" when dealing with difficult employees and shop stewards. Instead of ensuring that all the changes were enforced and normalised, they were sidestepped. Therefore, in a short time, the sullen shop steward team saw their opportunity and began to behave like they did before.

Some other ill-informed changes were introduced. A Board of Directors was put in place, consisting of the Plant Manager as well as the heads of HR, Finance, and a combined Quality and

Engineering Department. Overnight, this added another level of management, which went against all the modern trends in the industry. To make matters worse, the Board ceased to dine in the canteen that, up until then, the entire workforce had been using. Instead, they had their food transported from the canteen to the conference room through the factory floor. They stopped using the car park and employee entrance and chose to park at the other end of the building so they could enter without going near the factory floor. Readers can imagine the effect that such changes had on workforce optics.

Then, a very radical change was introduced by the new Board without any input from the broader management group. It was decided that we were to create Business Groups. Similar products would be grouped and managed by an independent team consisting of all the disciplines necessary.

This was a good idea and taken from an experiment carried out over some years by a large auto group. Nowadays, these are called Value Stream Organisations (VSOs), and I have been implementing them for many years as one of their biggest supporters. However, while I later led successful VSOs on three continents over the course of 20 years, it was only because of what I learned at Packard Electric.

The implementation at Packard was poorly planned, badly communicated, and executed in a perfect example of how not to do something. Porto cabins were put in place so that all disciplines could sit together. Engineering, Materials Control, Production Planning, and Quality Control staff were transferred from their central functions. All this happened within a few weeks or so, and no forward training of consequence was provided. No one knew what this would mean nor how they would operate. Some of the central function heads, although ostensibly supporting it, were suspicious of it, felt threatened by it, and took the opportunity to send their less-experienced and weaker people to the new teams.

What I have learned is that you must carefully put together the communications process intended for decentralised workers. It is easy for a young engineer or planner to feel abandoned by their central function and their roles misunderstood by the manager running the VSO. It certainly happened at Packard, where no thought was given to how central functions would relate to those they had transferred to. It is critical that those chosen for the VSOs continue to have a strong relationship with their central function (CF). They should continue to attend the CF department meetings, because people get strength and guidance from those around them performing similar functions. Moreover, employee appraisals and career planning should be conducted jointly in both their CF and VSO.

For that reason, I recommend that reporting lines do not change in the first year, so that those transferring still have that feeling of security and belonging. After a year, reporting lines back to the CF become a dotted line, and if the process has been working well, the kind of reporting line doesn't really matter.

At Packard, another new grade of manager was created: a Business Group Manager. A couple of superintendents were subsumed into the new roles, and the rest were promoted from the supervisory ranks. Now we had a new level of management in addition to the Board of Directors and the existing department heads. So, instead of trying to make the organisation flatter, we were adding levels.

Nowadays, I take months of careful planning, discussion, workshops, and coaching when steering a site to such a change. It will bring terrific benefits when done well, but it can cause confusion, demoralisation, and deterioration in performance when that kind of planning does not occur.

Suddenly, right around the time of the Business Groups implementation, the HR Head, Ronald, announced he was leaving. He had only been there about two years and had been the architect of the strike strategy and these other major changes. While he should be credited with being fearless in implementing radical change, he attracted much criticism for his policy of having us "walk on eggshells" and how, following the strike, he rid the plant of some of its main troublemakers.

A number of disruptive employees left the plant with handsome packages, which enabled some to start their own businesses. Although the benefits of this, on the face of it, were considerable to the site, it caused cynicism among many employees at all levels who saw bad behaviour being handsomely rewarded. However, Ronald had also injected much new thinking into the organisation, and many were sorry to see him go, especially considering who replaced him.

His successor was more self-serving than anyone who had been visited upon us until then. Somehow, this man had impressed the Germans at HQ, and he was quickly given an expanded role in Europe. As soon as he was hired, he advertised for a secretary. Fluency in German was a requirement, as were word processing skills (these were pre-laptop days). But he hired someone who had neither.

She turned out to be his *girlfriend*. He split his office in two so she could have her own. He then hired an admin to work for her, who did what his girlfriend had been hired to do. Those of us who cared about the site and worried for its future looked on, appalled.

In the meantime, I was getting on the best I could with the Materials Department. I was doing reasonably well with poor systems, though I did not truly realise *how* poor they were until I moved on to another organisation.

The issues at the factory seemed to grow as the years went on. Our financial monthly plant reports—reports that we should have been relying on—were woeful. The people who knew what was going on at the factory floor didn't believe them, and neither did most of those involved in producing the reports themselves. They were not widely shared and, apart from having summaries dropped on our desks and being asked to explain the deficits, there was no organised structure where real information sharing was taking place.

Most departments produced their own local reports and relied on those instead. However, Finance was the department most responsible for producing reports that would go to HQ and decide our future. I cannot comment directly on what they contained because I never saw them, but they would have been in a format approved by HQ and would have contained the basic information required.

However, I have long since learned that a factory needs much more than what will satisfy HQ or the local Finance Department. They seek summaries and bottom lines and trends, but on a local level, we need information that tells us how individual work centres are performing, and where the real costs are in raw materials, direct/indirect labour, and overheads. We needed to zero in on those, but at Packard, Finance had costs rolled into large bundles, so it was only convenient for them. This made it difficult for us to see where the opportunities were.

Finance leadership were without vision, had little or no understanding of manufacturing, and because they wielded such power, they had an arrogance that belied their incompetence. We had a budgeting system that was almost non-existent and controlled by a tiny number of people. The reports we received each month were error-ridden and useless when trying to extract useful trends. It would be later, when I moved to Oral-B, that I would learn how a forward and open-thinking finance team, with a well-integrated set of reports, can both inform and lead a plant to continuous improvement. But this was not the case here.

Similarly, our warehouse was causing many problems. It was now the late 80's and space was an ongoing issue. New quality standards demanded levels of segregation that we were unable to provide. So, I was granted a project, along with $300k, to transform our warehouse into something that could accommodate all our needs and improve support to the factory floor.

This was my big opportunity to really do something that would change how we managed our materials and supported production. All the learning I had been devouring at the Irish Management Institute, where much new thinking was being taught, was going to be brought to life. The books I had read, as well as the other sites I had visited and learned from, were going to be applied.

I put together the first ever cross-functional team on that site, developed a mission, scope, and calendar, and presented it to the Board for approval. I said we would report to them once a month, or outside of that if we had an emergency. Otherwise, we would just get on with things. I deliberately set it up this way because I wanted to keep department interference to a minimum. I knew some would resent our independence.

The project team consisted of myself, Noel, who was a genius with numbers and Excel (in 1988 it was still quite a new tool and not yet widely used), and our Warehouse Team Leader, Charlie, a no-nonsense, well-respected guy who was trusted by his people and could get things done. This was essential, because I knew in advance that many work practices would be changing, and we did not want the unions to get involved and disrupt that. We were also allocated an experienced project engineer and two young engineering interns.

We had a couple of team meetings where we went through how we were going to operate and communicate. It was important to me that everyone would have a voice and be listened to.

One of the first things we did was engage with a third-party designer and supplier of storage systems. As we had a relatively small sum of money for the project, we figured if we committed to giving them the business for storage systems and fork trucks when we decided what we wanted, they would provide whatever support we needed while working through the details. This was committed to with the stipulation that we would check the marketplace to ensure they were being competitive.

This worked extremely well, and they took us to see several sites, including one in England that they had designed and installed. Like Packard, it was also a manufacturing plant, which made comparisons easier.

At this period in our factory, one of the most time-consuming tasks in the warehouse was supplying large reels of cables to Production. This required someone on a forklift, on every shift, selecting and transporting reels of cables to the factory floor and bringing partially consumed ones back. With an increasingly busy factory, this activity was not just expensive, but dangerous. We had several minor accidents and near misses by people getting hit by forklifts or having a reel of cable dropped on them. It also devoured a lot of increasingly valuable floor space and a huge tract of our warehouse storage.

So, my team came up with the suggestion to reallocate the cutting machines consuming the cable and asked that we supply the cable through holes in the wall instead. If this was accepted, we knew it would be a brilliant development. We could already see the cost savings that were possible.

We then commenced discussions with the cable and other key suppliers to arrange for Kanban deliveries. We established direct links from our stores personnel to trigger direct signals to our suppliers when replenishment was needed. We had min/max volumes for each item clearly marked on the storage racks, and a map where each item was stored. Each warehouse team member had their own area for which they were responsible. I had seen similar systems work well on a site in Portugal.

We also designed a mezzanine floor in the warehouse based on supermarket principles. Stores personnel were given specially designed trolleys for families of products that would roll in and out of the Production workstations. They didn't need forklift trucks; they would walk around, pick the materials, and go straight to the production lines. Forklift trucks were banned from the shop floor as they were no longer required. This was quite brilliant in its simplicity, and we could immediately see that this would not just transform how our warehouse operated but would also deliver a new manufacturing system.

Before proceeding with the concept, we needed approval from our Plant Manager as well as HQ in Germany. This was such a radical change that it was difficult to communicate by drawings alone, so we decided to have a model built that would clearly show what we were proposing. We found a young local company who were trying to get off the ground, and they produced something quite brilliant that excited everyone about the change. I brought it to Germany and got their full support, so we were on our way.

We rented a temporary warehouse just 200 yards from our site and we moved all our materials there while we went to work dismantling and assembling our new creation. In just 12 weeks it was complete. It overdelivered on every front. The results were truly impressive:

- 70% of factory floor space was freed up
- 40% of warehouse space was freed up
- Raw material turns went from 7 to 24
- Finished goods inventory turns increased from 6 to 22
- Schedule reliability to our customers increased from the low 80's to 98+%
- Cost savings in excess of 2 million pounds

Overnight, service to production was transformed and provided on a Kanban basis. The new system brought stability and calmness to our factory. Previously, many workers were caught up in chasing materials and arranging emergency deliveries and changeovers. That all but disappeared. We stopped machines running product that was not required. The environment had been transformed.

There was just one unfortunate issue.

Our engineering department had a close and long-term relationship with a supplier that some felt was *too* close. This supplier had provided previous storage systems for our cable, and just as we were about to confirm an order with the third-party storage designer we had entered into an agreement with, the Head of Engineering announced that they had carried out tests on the third party's proposed storage racking and it did not meet the safety specifications required. In the new layout, cable would be stored at heights higher than what we used before, so this was the primary concern. To no one's surprise, they were recommending their regular supplier.

I objected strongly, as we had a written agreement with our chosen designer that everyone had been made aware of. Furthermore, information concerning the test was sketchy, to say the least. Engineering had an experienced senior project engineer on the team. He had been part of all our team meetings and had helped us make every engineering-related decision. He had visited sites to view similar storage systems. However, he was on a temporary contract with Packard and was fearful that, if he crossed his boss in Engineering, he would not be around for long.

Still, I maintained my objection to using this regular supplier. So, Kurt, the Plant Manager, said he would hold an "inquiry," but in the same breath, he also told me that another director would take the lead so it would be "independent." However, this person was also close to the supplier and the Engineering Department Head, so I knew it would go nowhere. As I expected, the contract went to the regular supplier.

It was a travesty and an injustice to me and the rest of the team. We had let down our third party who had performed quite brilliantly for us.

However, it must be said that I had no issue with the existing supplier. They were competent, flexible, and, overall, provided an excellent service. In fact, when I moved on, I introduced them to Oral-B and used their services during my years there.

But I still felt we had turned our backs on our third-party contractor. This was another lesson learned. In a nasty political environment, like what existed at Packard, when you think you have covered all the angles, have another look.

A short time later, after I had moved on to Oral-B, I got a call from my old warehouse team leader. He said that a portion of the storage racking purchased from our regular supplier had collapsed the previous day. Thankfully, no-one was hurt. But it just reinforced the sense of injustice that I and my team members had felt at the time.

Meanwhile, Manuel, the new Head of Manufacturing, was

brought in on assignment from a plant in Portugal. He was a nice guy and well-intentioned but completely out of his depth in our environment. Typical to the experience at Packard at the time, he was not well supported by his colleagues in management who saw him as a threat to their ambitions.

After some months with production performance struggling, Kurt asked me to leave the Materials Department and go back to support Manuel in Production. Ostensibly, I was being asked to develop the Business Groups (VSOs) that had already been established a year previously. But the real reason was to support Manuel.

Initially, I resisted. I had just led a transformation of our entire supply chain. We had the most admired system not just in Europe, but in the entire world of GM, and I was on top of that world. I was getting calls from all over. People were coming every week to see it. I was being invited to sites all over the world to come and advise them on how to improve their operations. My old boss, Jim, who was now back in the USA, asked me if I would team up with him, become a consultant, and earn a fortune in the States! Moreover, a program at the Irish Management Institute adopted it as a Case History in their curriculum.

I would be lying if I said I was not enjoying it. I really wanted another year or two in Materials Management to polish some things and to enjoy the terrific results we were getting.

But Kurt persisted, and my sense of loyalty to the plant overcame those feelings, so I became the Business Leader Development Manager. Sadly, it was a move that did not result in that loyalty being reciprocated.

I threw myself into the role the only way I knew how. I helped develop a range of programs for the various teams and individuals. We used external coaches to teach them how to communicate better. Supervisors were taught how to solve problems by role play. Shift handovers were formalised to make them more effective. Existing reports were either scrapped or redesigned. I organised a couple off-site weekends to help build confidence and teamwork. I provided Manuel as much support directly and indirectly as I could, and things were moving along reasonably well; results were improving.

But then, less than six months into the role, Manuel went home to Portugal, and a new Head of Manufacturing was recruited. We had many poor appointments at that site, but this was up with the very worst.

It was then that I learned mediocrity will always seek out mediocrity, or worse. A mediocre person will not hire those who will threaten them or their job. Several senior people had ambitions to lead the site, and they did not want to hire someone who would threaten that.

The new Head of Manufacturing, Patrick, was paranoid and had exaggerated his background. Even cursory checks would have exposed that. He was suspicious of anyone who might have posed a threat to him, and he did not understand modern manufacturing thinking. He could not comprehend the concept of the Business Groups, or Kanban for that matter, and over two decades later, it is still quite astonishing to me how someone like him could get a position like that.

Within weeks of Patrick's hiring, my role was being isolated. He saw me as a major threat, and I found myself with less and less to do. I discussed it with Kurt, and he blamed Oliver—the same HR guy who had hired his girlfriend and was now spending much of his time on business trips with her. I didn't buy Kurt's explanation—both were responsible.

So, I was offered a role in Customer Relations & Development. But after a couple of months, this had still not been finalised, and it would have included a lot of travel. With three young children, I was not at all convinced this was the right thing for me.

A few months previously, I had turned down an offer from Oral-B Laboratories. Their factory was based in Newbridge, Co. Kildare and conveniently close to where I lived.

So, when they called again, I accepted the offer and handed in my resignation to GM. I had been with GM for 21 years, and though I asked for a redundancy deal, I was denied. Many people, including a lawyer friend, advised me to sue. I would have received a handsome settlement for sure, but some months into my new role at Oral-B, I became happy with my decision and saw the potential there. So, I put GM behind me.

To this day, I believe I am the only senior manager with over twenty years of service that left GM without a package. I am quite proud of that! A couple of years ago, I was speaking at an auto conference in Nashville and, as part of my introduction, I told them the story. I was surprised how many came to me later to discuss just how surprised they were. Attrition rates tend to be high in the auto world, but they come with handsome separation packages for long-serving people.

Despite the unhappy ending, I remain enormously grateful for all that I learned with GM. I joined as a young man shortly after leaving Smiths, unsure of what the future held. They groomed and educated me, promoted me many times, and gave me plenty of opportunities. The lessons in manufacturing I received there put me in a strong position to move on and bring what I had learned to other levels.

I also learned how *not* to do things. I saw behaviour from senior staff that was unacceptable. I saw incompetence that defied common sense. I saw managers at every level put their self-interests above the organisation and the employees they were responsible for.

Six years after I left, I was in my car when news came over the radio confirming that the GM plant was being closed. Although it came as no surprise, I still had to pull over for a few minutes. It was an emotional moment. When one spends so long in an organisation, you leave something behind you. You cannot erase all the feelings and emotions. I thought of all the great people that were still there, the damage caused by inept and self-serving leaders, as well as the arrogant, small-minded union activists who had contributed to its demise. To this day, I am convinced that, with the right leadership, the plant could have evolved and survived.

Lessons learned at Packard (GM) that I carried with me for the rest of my life:

> - A divided leadership team will not be successful in developing an organisation
> - Disloyalty and incompetence should not be tolerated and should be dealt with quickly
> - WCM (Lean) Practices will provide outstanding results
> - Most production problems originate away from the factory floor
> - Mediocrity will seek out mediocrity, or worse.
> - Do not allow union reps to come between you and your employees
> - Ensure sound, two-way communications processes are in place so that everyone has a voice that is both listened and responded to
> - Establishing Business Groups (VSOs) requires a lot of careful communication and training, as well as the design of appropriate processes

CHAPTER FIVE

Case Study: Oral-B Newbridge
A Division of Gillette

Many of my friends advised me against leaving GM. They said after being with them for so long, I should have waited for something else to come down the track. But I'd had enough.

Even though it gave her sleepless nights, my wife supported me in my decision, and that mattered to me the most. She had seen how much I had given to GM and how it was negatively impacting me. But she wasn't the only one not sleeping—I was going into a completely different industry, and I didn't know a soul at my new job other than my interviewer.

The Oral-B plant was established in 1984 with just 30,000 square feet of space, but it had a couple of extensions, including one before and another after I arrived, bringing it to just over 110,000. It eventually grew to 400 employees.

Oral-B had a separate Good Manufacturing Plant (GMP) where medical products and a range of creams and shampoos were produced. It was a modern site on the outskirts of Newbridge Town, and it was close to the famous Curragh racecourse which attracts many thousands of visitors during the flat racing season. Large studs, where horses are raised and trained, are scattered all over that region. Newbridge Town also borders on what was once the largest military site in Europe, containing a total of seven barracks all adjoining each other.

So, the primary employers in the area were the military, the horse industry, and manufacturing. The other large manufacturers were the famous Curragh Carpets and, later, Wyeth, the giant pharmaceutical that was later to be acquired by Pfizer.

As the Manufacturing Manager, I was hired to introduce World Class Manufacturing (Lean) to the Oral-B site. Kevin, the Managing Director, was a dynamo of a man who had built a campus in less than a decade. He had incredible energy and bullied and cajoled his masters in California to transfer various products to Newbridge, which made the site more critical to the wider organisation.

Kevin could be both terrific and difficult to work with. He needed something new to be working on, otherwise he got bored, restless, and difficult. He was unreasonably demanding at times, and many people could not take it and moved on. However, the general workforce loved him, saw that he really cared for his site, and in an era of high unemployment and job scarcity, jobs were being created at Oral-B.

He deserves much credit. Kevin had a talent for spotting the kind of resources the site needed as it evolved, and he built a reputation for developing people. Some of his protégés, including myself, became leaders at other Gillette sites around the world, and many others who left the organisation rose to senior positions elsewhere. Working under him, the leaders of tomorrow cut their teeth.

In short, he was well-suited to the role of establishing and growing an operation. He needed the adrenalin of those big new projects.

However, when the plant became more established and needed stability and better systems, Kevin found it difficult to adjust to a regular, day-to-day routine. He thrived on major change.

When I arrived, I found a much different environment from the one I left at GM. The leadership team was united for the common good. Most struggled at times with Kevin's demands, but all recognised the strengths he brought to the table. The workforce just wanted to get on with things, and union disputes were almost non-existent. This was due, in no small part, to the role and environment the site leader had created.

Kevin personally toured the site at shift change every morning. He would comment on things he saw, sometimes being complimentary, and just as often being critical. However, he rarely criticised the shop floor personnel, usually reserving his comments for supervisors or managers.

He also developed what are now known as Town Hall Meetings. Every quarter, the workforce would assemble in groups, and results, accomplishments, and issues were presented and discussed. The results also determined the level of bonus employees would receive. He had introduced a plant-level bonus system that was dependant on how it did against its goals, and while the system was not perfect, it was ahead of what most organisations were doing at the time.

Something else that was ahead of its time: when problems existed, he would insist on including the semi-skilled technicians and operators in finding a solution. Most organisations didn't include their shop floor personnel in these decisions. Their opinions may have been sought, but it was rare to see them brought to a meeting room to have an equal voice in working through the problem.

He was also instrumental in initiating a cross-craft training program with a local 3rd Level Institute, whereby electricians and mechanics learned each other's key skills. This was the first of its kind in Ireland and led to many other organisations developing similar programs. It resulted in significant improvements in response time to production issues.

Kevin also had a process of collective input when preparing ratings for annual appraisals. Each department head was encouraged to seek input from their teams, and we would then meet and go through each staff member and discuss which rating was appropriate.

Basically, the Oral-B system had three ratings: Highly Effective, Effective, and Needs Improvement. At Packard, decisions like this were left to the individual department head, which produced uneven and unfair results. Some department heads had almost everyone at Highly Effective, while others insisted most of their staff needed improvement. It caused a lot of unrest. When a formal process tells you what your employer thinks of you, it is serious because it can impact salary increases, bonuses, and promotions.

So, I found Kevin's process to be a much-improved way of providing ratings, because it included every department head's input. Moreover, before meeting we ensured we had input from our middle and front-line managers, so we were clear on what the recommendation should be. I have carried that practice with me ever since.

Kevin had created a strong foundation to build upon, which was fantastic for me. I came into a good environment, and I had gained much confidence from my successful project at GM. I was confident I could make a strong contribution.

The Oral-B and Gillette culture was so much more attractive than GM. In those days, the auto culture was relentless, nasty, and very survival of the fittest. I saw it break many people. At Oral-B Gillette there was a real sense that, as an organisation, they wanted to embrace change and care for and develop their workforce. We had often joked, half seriously, that if you made it to forty at GM without going mad or worse, you could survive anything! It was such a pleasure to leave that behind, and the conditions of employment were so much better at Oral-B. Up and down the organisation there was a potential to grow.

Soon after I began, we started to roll out the Lean Tools. This was relatively easy because we had a willing workforce who just wanted to do their best, and if a new way made sense to them, all the better. We introduced the recording of key data at all work centres. These were visual to everyone, and we could enter into a discussion with individuals or teams about their results and issues. Operators would record their own data, including output vs targets, scrap levels, quality or technical issues, and reasons for lost time. It was simple, but this basic information is as important today as it was then. Organisations don't need expensive systems to do this.

Before we commenced recording data manually at each work centre, we would receive monthly financial reports based on what I refer to as "rolled up" summary reports. By "rolled up" I mean clusters of machines are joined together to make a single work centre.

While at times this does make sense, especially for low value products, I much prefer to have individual machines treated as work centres. This way the data is much easier to analyse. Systems like the "rolled up" kind are usually driven by Finance with their needs in mind, which is what is most convenient for them. However, at Oral-B, when we compared our local manual data, it bore no comparison to what Finance were producing. Their waste percentages were much higher than ours. When we dug deeper, we discovered that our Bills of Material (BOMs) were highly inaccurate. They were being maintained by a junior accountant, and although well intentioned, when he was updating or adding new items, all kinds of assumptions were being made about similarity of materials, rather than checking each individual item.

Still, compared to GM, the Finance Department was well led, and with reasoned argument, they were open to new thinking.

But eventually, we took this responsibility away from Finance and gave it to someone in our Quality Department. They spent six months going through each of our BOMs. This transformed our financial reports. They became similar to our own manual results, and therefore, useful.

I have found this same problem in almost every site I have worked on. Too many organisations simply do not put the resources into maintaining databases, and the result is that the database cannot be trusted. My advice to clients is to clearly understand what resources it takes to maintain these databases, and do not compromise on that. Preferably have that discussion before you invest in a new system, because it is not a point that will be emphasised by those selling database technology.

Data does not appear as if by magic. Its source is a database that needs maintaining just like your equipment does. I am fearful of the rush many organisations are taking towards Industry 4.0 without understanding the levels of discipline and maintenance it will require. I will comment on this later.

At Oral-B, we also introduced Kanban deliveries from warehouse to the factory floor, and later implemented Kanban on Finished Product to customers in the UK, Germany, and Italy. We concentrated on fast changeovers, later to become known as SMED. We phased out supervisors and replaced them with hands-on Group Leaders. As there was only a small number, they evolved into shift managers and spent much time supporting projects and embedding our changes.

In 1996, I wrote a paper for the Institute of Logistics. It was called 'Transforming a Manufacturing Organisation,' and it summarised what we had done at the Newbridge site. It led to two awards in two years—one for 'International Logistics' for improving inventories and supply for our European customers, and one for 'Best Paper' at their annual conference. It helped put our plant firmly on the map within the wider organisation.

The awards were presented at the Dorchester, Park Lane in London, and one of them by Her Royal Highness Princess Anne, who was the patron of the institute.

Her Royal Highness Princess Anne, Patron of the Institute, presenting the International Logistics Award, London 1997

Coming from a border county, and with our Irish history with British Royalty, the latter award would not have impressed too many from my community, so I said little about it. However, it was nice to go to Park Lane—one of the most famous streets in the world—to receive it.

I include the paper below. This, in part, is what my team and I, with terrific support from the site leader and the rest of the leadership team, accomplished in my first three years at Oral-B. In particular, I should point out that HR were effective in helping implement the changes. We were fortunate to have forward-seeing leaders in that role who encouraged and facilitated the Change program.

There was a simplicity to our approach. We didn't make any grand announcements, thereby alerting unions to the possibility of making claims for additional payments. We knew we needed to do this to secure our future, and we started off in small steps, involving only those immediately impacted, until confidence and acceptance grew, and we were able to pick up pace. Change can be frightening for some, but keeping things simple makes it easy for everyone to understand.

For example, there are formulas for calculating Kanban min/max numbers and readers can find them online. However, I asked those people in the areas impacted to come up with recommendations that, from their experience, they could operate with. I knew they would initially be on the high size, but once they got familiar with it, they would seek reductions. This is exactly what happened.

Far too much "Change" is shrouded in jargon and mystery. Strip it away and what you will see is mostly common sense!

It should be noted that this was written in 1996. Much of the language has changed since then, and indeed some of my approach has too, as you will see in further case histories. With practice we learn and get better!

Transforming a Manufacturing Organisation

In 1992, there were less than 200 Stock Keeping Units (SKUs) which allowed lengthy production runs—anything from two to eight days. The system was a mixture of flow (mass) and batch manufacture driven by Materials Requirement Planning (MRP).

Each production department had its own separate distinct plan and utilisation of equipment and maximisation of output were key measurements.

The traditional problems associated with such a system are already well documented. Oral-B was experiencing so many of them. Problems such as:

- High Work in Progress (WIP): up to 30% of floor space occupied
- Frequent material shortages, despite high value inventories – items also frequently getting lost
- Inaccurate inventory reports
- Changeovers frequently not occurring when they should
- Growing number of 'past dues' to our customers
- Inaccurate Bills of Materials (BOMs)
- Long lead-times
- High inventories in regional warehouses
- Erratic customer forecasts

We had already implemented elements of Manufacturing Resource Planning (MRPII) and whilst this had improved the ability to monitor individual work centres, we were unhappy with the additional layers of bureaucracy and massive increase in transactions which it had required. There were volumes of reports available on a daily and weekly basis, but in reality, they were virtually useless.

Manufacturing is complex, particularly in batch production, and we were already experiencing serious problems. It was clear also that significant changes were going to occur in the marketplace: product life cycles would get shorter; there would be an increase in SKUs; lead-times would have to improve; a significant increase in batch production would be expected.

In other words, faster response to customer demands would be mandatory. It would not be possible to resolve the problems or improve the systems by adding further layers of complexity. Rather, a much different and more radical approach would be required.

It was decided to introduce a programme of World Class Manufacturing (WCM) which would include the philosophies of Just-In-Time (JIT), Continuous Improvement, Customer Partnership, and techniques such as Kanban. We realised that this would require a significant cultural change and a considerable investment in preparation, education, and implementation. However, we also realised that if we were going to position ourselves to not just survive, but to take advantage of the rapidly changing marketplace, that change was inevitable.

Our workforce was young and committed, and, we believed, receptive to change. The exception was our craft group, about 15 strong, who were traditional in their outlook and orientated towards their union. They were suspicious and resistant to change and opposed to any integration with the production groups. Any changes contemplated here would require a specific approach.

Planning and Preparing

We were determined to take a pragmatic approach to the change programme. We wanted no fanfares, no hype, and no unnecessary heightening of expectations. On a more practical level we wanted no grading claims, as fundamental to our philosophy was the fact that on-going change was necessary to survive. We were also determined to take a practical view of the many versions of how such programmes could be implemented. We would do this ourselves! We would not use consultants. We would not become slaves to any of the many gurus that existed. More importantly for us was to have a good understanding of the philosophy and the techniques contained within it and how best they could be tailored to our requirements.

As we viewed it, the essential requirements to get the programme up and running existed. We had:

- Commitment – senior management was fully committed
- Education – would be on-going for the workforce; we already had a monthly forum of team meetings which we would use
- Key project team members – sufficiently experienced people existed; this would be supported by benchmarking and seminars
- Defined objectives – we would ensure that they were clearly understood and measurable

This, then, was the basis for the change programme.

Setting Objectives

Again, practical considerations largely determined what the objectives would be, and in which order they would be implemented. We avoided saying things like: 'We want to create a new culture.' Instead, we said things like: 'We want to reduce space allocated to WIP by 20% to create space for new processes.' And: 'We want to improve our lead times to our customers.'

In other words, we avoided using much of the modern jargon that often means little and is rarely understood—try cross-examining the person using the jargon sometimes! In practical terms the objectives were:

- General introduction of WCM (Lean) techniques to the workforce
- Introduction of Kanban from warehouse to factory floor for all raw materials
- Movement within factory floor would also be controlled by Kanban – with sensible exceptions
- Clear separation of MRP from production schedules – must be visible and controlled by shop floor personnel
- Develop 'Mixed Modelling' planning system
- One overall production plan for all stages of process – level load
- Reduction of floor space allocated to WIP reduced from 30%-10%
- Introduce Cross Craft Training Programme
- Re-engineer production process
- Re-organise engineering spare parts supply
- Commence Supplier Development programme
- Develop training programme for our customer's logistics personnel
- Analysis of total supply-chain
- Introduce 'Pit-Stop' changeover programme
- Introduce Kanban on 'Star' products to international customers
- Give visibility to European inventories
- Introduce 'Value' teams – self-directed

Implementation

Kanban

The first schedule was put together in early 1992. A relatively modest one, the main objective was to commence implementing Kanban from the warehouse to the factory floor. We had put together our project team on the basis that it would have two permanent members who would move from area to area, using key local personnel for implementation in their own departments. That way adherence to the overall objectives was assured and local detail—and ownership—was provided by the local members.

The first area chosen was the Injection Moulding Department. It consumed only a small range of materials and demand was pretty regular. In other words, it provided a very huge likelihood of success—a crucial requirement of the team.

In the physical implementation of Kanban, the real key to success is attention to detail. The philosophy: 'A place for everything, everything in its place', lies at the heart of it. In this was everything, that was equipment—including ancillary systems often omitted from layout drawings—all material, miscellaneous requirements such as lockers, holders for charts and markers are captured. A defined space is identified. If something is not in its place, then it becomes immediately apparent.

Therefore, ensuring everything is included on the layout is crucial. Defining minimum and maximum quantities of materials, and the cycle of time replenishment, is where time has to be spent to ensure that it matches correctly. Local conditions will determine many of them. For example, why replenish every eight hours if cost and space is not prohibitive? Why not every 24 hours, or 48 hours?

That way changeovers and material handling costs are kept in balance. This is where local operator knowledge is important. I do not believe you can successfully implement without their active involvement.

Determining 'a place for everything' and replenishment cycle times requires knowledge which often only exists at that level. We made sure they were involved and gave them reasonable time and scope to articulate their feelings.

In less than six weeks, Kanban was up and running in our first area. Large areas of floor space were freed up and overnight we stopped running out of materials. The frustrating kinds of communication that had existed between warehouse and mould personnel ceased. It was seen to be an outstanding success. We were on our way!

It is not possible to go through all the detail involved in the implementation of the entire Kanban programme. What I have attempted to do is to illustrate how we prepared for it and how we set about it by having a mix of permanent and local team members involved in developing the details and its implementation. Once started, it actually got easier. Even in areas where more complex processes existed, difficulties were quickly overcome.

Enthusiasm grew so quickly that at times we had a problem slowing the programme down to ensure that essential detail was not being missed.

Before the end of the year, the entire shop floor, inclusive of all movement from the warehouse, was controlled by Kanban. Floor space previously occupied by WIP—up to 30%, was reduced to 5%. We had targeted 10% and had therefore overachieved. WIP inventories also reduced dramatically.

Stockouts, materials 'lost', emergencies and a number of other activities—meetings, memos and reports—have virtually disappeared. Technical difficulties that arise with Kanban have all been overcome. For example, if a bottleneck exists due to a breakdown, Kanban is simply suspended—special cards were made up for this. Once minimum stocks are replaced, suspension is lifted. If a capacity storage exists, we use rate-based scheduling—but we retain visibility by using a variety of cards and defined locations. The principle of visibility and 'a place for everything' is deeply embedded.

Implementation Supply-Chain Analysis

We carried out an analysis of our supply chain. We identified bottlenecks, many of them administrative, and improved or eliminated them. We sent our customers' freight related questionnaires and phased out a number of contractors who could not improve.

We established Kanban services on a range of our 'Star' products with our main customers in Europe. Although traditionally seen as an internal tool, we saw no reason why a signal from a warehouse in Italy direct to our own warehouse calling for a replacement product was that much different from a signal from one production department to another. The principle is similar, the detail is different. Pay sufficient attention to the detail, get it right and it will work.

Culture

We modified Richard Schonberger's model of how WCM (Lean) works. Our select focus was to move from being customer-driven to customer-partnership. The key factors of customer service are:

- Responsiveness
- Reliability
- Quality
- Courtesy
- Price
- Problem solving

We focused, particularly, on two gaps at the provider/customer interface:

- The gap between customers' expectations and the provider's understanding of those expectations
- The gap between the customers' memories and experiences of what was delivered and the providers

These gaps occur not only in actual outcomes, such as product delivery, but also in communications.

Used internally—it was included as an element of all staff appraisals—as well with our external customers—it changed our views on our role as a manufacturer. It was one of the most significant contributors to the cultural change that occurred.

Level Load Planning

The principles of 'level load' planning and 'one' production plan were introduced—leading to equipment being switched off when consumption slowed down or ceased—a practice previously unthinkable.

This was the biggest internal obstacle encountered. Many people found it difficult to adjust. In the early days of the programme, some supervisors and operators were so frightened of running out of material that they attempted to run parallel systems to avoid switching off. Of course, with space defined for everything, this quickly became apparent and was easily dealt with.

An internal training programme for our customer's logistics personnel was developed and had led to much improved understanding of the various forecasting and planning systems. European inventories have been halved!

Cross Craft Training

Meanwhile, many other things were happening throughout the plant. Negotiations were concluded with the Craft Group and a two-year certified cross-training programme, the first of its kind in Ireland, was concluded. Key aspects of this programme were decentralisation and integration with the local production teams.

Today, the behaviours of our craft members are totally changed from what it was prior to 1992. They have workstations located next to their production team; they attend the monthly team meetings; they are actively involved with the local team—their team—in improvement programmes.

Many of them are continuing their education, availing of another in-house course in mechanical engineering which we have negotiated with a local college. We are confident that some of them will become fully qualified engineers… a far cry from our previous views of many of them!

Quick Changeovers - SMED

Using the racing car 'Pit Stop' approach, we developed teams in each department. In practically every case, initial study indicated 'walking time' as the biggest contributor to lost time. In other words, poor preparation time resulted in personnel constantly going to have to look for spanners, gauges, and drawings. By itemising each item required for each type of change and allocating tasks to A, B, and C resulted, in some cases, to hours being taken out of changeovers.

Outsourcing

We saw that fluctuations in the marketplace, that would lead to sharp movement up or down in our plant schedules, would continue, despite the various improvements that were being made. This was resulting in large swings from month to month making it difficult to retain stability on headcount and materials.

In the supply-chain analysis we had identified opportunities to both remove steps from the chain and also reduce inventories. Some of our customers were applying 'secondary' packaging for specific customers. They were also doing a considerable amount of re-packaging for promotions.

Much of this, we felt, could be done at the manufacturing point in Ireland. However, the lack of space and erratic nature of the demand made it difficult to handle in-house. We decided that if we could successfully outsource, we would overcome, what, for us, was the serious problem of having a stable headcount and at the same time offer a more cost effective and improved service to our customers.

We already had a relationship, albeit a small one, with an organisation which specialised in providing training for handicapped people. They also supplemented their activities with fully-fledged industrial units. We felt that, given reasonable supports, they could be developed to provide us with this service. A number of other factors existed. They had a number of sites—all within 15 miles of our factory. They were anxious to develop into more complex processes. Much of their activities were very low skilled and were not allowing them to challenge and develop their people.

We commenced with some small operations, gradually building it up as their skills increased. We appointed one of our supervisors as their facilitator, someone who would support them in training, layout, design, basic work methods, and material control, including Kanban.

Today, we have a dynamic operation, operating on three sites and employing almost 100 people on our projects. It has allowed us to overcome the difficulties referred to earlier in relation to headcount stability and lack of floor space. It has also enabled us to perform operations previously carried out by our customers at reduced cost; thereby improving our all-round service.

Value Teams

We have empowered teams throughout the plant to take over large amounts of the tasks previously carried out by supervisors. Changeovers, overtime control, emergency absence, housekeeping, and safety audits are now handled on a day-to-day basis by the local teams.

Our skilled and semi-skilled people are regularly in touch with suppliers on technical issues. In one of our main departments, we have reduced the supervisory cover from one a shift to one covering both shifts. Supervisors have not been made redundant—we value their local knowledge too much for that. They now work on improvement projects, new products, and improving safety through audits and safer practices.

Project Management

A plant-wide system was developed to manage and control projects, referred to as ProMPT. It means Projects Managed and Performed on Time. It is a basic tool with a common language and structure to help thinking. It is PC networked with total access by all participants.

Linked to an adapted version of Microsoft's Project management programme, a common way of doing new things, while assuring company priorities determine resource allocation, is actioned. Communication is live through the network and a leader, clearly responsible, is always specified, whether the project is big or small.

It has opened up project management to many people who previously would not have been given formal projects—such as administrators, supervisors and team leaders. We have also recently expanded it to some shop floor personnel.

The Results

Our manufacturing system today processes almost 1000 SKUs each month, has same-month lead times or, at worst, the following month. Our system operates almost entirely on JIT/Kanban basis, and we have extended direct warehouse-to-warehouse links—which we still call Kanban—to some of our distributors in Europe.

The Kanban systems have also been extended to include the internal movement of over 2000 regular engineering parts and the replenishment of those by a range of suppliers, much like supermarkets have shelves restocked.

Cycle time, or response time, is now established as being focused towards customer satisfaction, while contributing to economic performance.

Key results include:

- Lead time from suppliers are shorter by 60% - in transit times and, more importantly, in response times from order placement
- Past dues were an important measure, but when we consistently achieved 99.5% reliability, we moved on to managing our customers' inventories in Europe – reduced from 3.1 months in 1992 to 1.7 in 1995
- New introductions here increased by 40%

In three years, the company has:

- Halved turnaround time twice
- Halved inventory twice
- Reduced costs by 10% twice
- Increased product offerings by 40%

Along the way:

- The company became certified to ISO 9002 and Environmental Management System BS7750

Was overall winner of:

- Achievement in Personnel Management – 1995
- International Logistics Award – 1996
- Transforming a Manufacturing Organisation Best Paper 1997
- Are previous holders of the Supreme National Award from the Irish Quality Association

In Conclusion

The changes implemented here have by far exceeded expectations. This article has touched on many, but by no means all of them. All over the factory individuals and teams are working on improvement projects and new processes with an energy and determination that comes from being both empowered and challenged.

WCM (Lean) has offered a new world of opportunity to our workforce—at every level. It has stimulated innovation and re-energised us to a degree that would have been difficult to imagine a few years ago.

It has been achieved by practical people taking a practical common-sense approach. We did not have expensive training programmes; we did not use consultants. We used what we believed to be our single biggest resource—the experience some of us had previously gained and the enthusiasm and skills of our own people.

What Next?

- Needs to be a question asked not once, but continuously
- Only those firms who willingly embrace change will be the winners in the race of competitive supremacy

- 'Who will survive?' 'Companies that adopt a constancy of purpose for quality, productivity and service, and who go about it with intelligence and perseverance, have a chance to survive.' – W. Edwards Deming.

Looking back at my experience during this time brings back many memories. I was fortunate to have the support of the site leader for the program of Change. He was demanding and had high standards, but Kevin asked for little that he was not prepared to do himself. I had a willing and open-minded workforce who accepted that, to survive, we had to implement changes and continuously strive to improve.

Moreover, I had the full support of my colleagues on the leadership team. I learned that to lead is to not make all the decisions, because while I was steering the program, I was also happy to delegate, step back, and allow the various teams to get on with things.

Throughout this process, I drew heavily on all the positive experiences I had gained at GM, but I also learned from mistakes we made there, and still do to this day. I frequently tell younger managers that, every year when I look back, there are some things I would do differently. Learning from experiences is never ending.

Six years after I joined Oral-B, Kevin accepted an assignment at HQ in California, and I was promoted to replace him. For the next two years, I streamlined the leadership team and continued to lead the development of the site, cementing its reputation within the wider organisation.

But change, often unexpected, is a constant in global organisations. In 2000, Oral-B merged with the German Braun division, and they were anxious to create a campus consisting of the Oral-B Sites in Newbridge, India, and the Braun factory in Carlow, 30 miles away. However, I was not chosen to lead the new campus. The existing Plant Manager in Carlow was selected instead, and I was asked to first go to Iowa, USA, then return to take over from him when my assignment there was complete.

This was a huge step. The site in Iowa was the largest toothbrush factory in the world. Back then, all Oral-B manual toothbrushes were designed at the HQ in California and product development took place in Iowa. This made it very strong technically, but on the manufacturing side it had not embraced new thinking. Other sites, including Newbridge, had been substantially reducing costs, but Iowa was not. Successive leadership had not embraced the changes that their sister factories and some competitors had, and the plant was dying. I was asked to prepare it for a closure announcement within two years.

But I had other ideas.

Lessons learned at Oral-B Newbridge that I carried with me for the rest of my life:

> A site leader that is visibly demanding of the leadership team will be trusted and admired by the workforce
> With right processes in place, workforces will respond and embrace change that will benefit both them and the organisation
> A united leadership team can accomplish incredible things
> Outsourcing can be used as a tool to free up space and have high labour operations produced in a lower-cost environment
> A consensus-driven process involving all department heads is an effective and fair way of allocating annual appraisal ratings

CHAPTER SIX

Case Study: Oral-B Iowa City Fighting for Our Future

My boss had joked, half seriously, that if he sent a German to Iowa to tell them their plant was going to be closed, he or she would be ran out of town. But, as I was Irish, I would get away with it!

Regardless, I had no intention of seeing the plant shut down; not without a fight anyway! I saw it as a great opportunity to put into practice what I had spent years learning. I had terrific success at Newbridge and Packard

106

Electric, and I felt I could build on what was a much larger site with great potential.

I was excited to move to the United States. Ireland has always had a special relationship with the USA. Most Irish families have relatives there, and almost 60 million of its citizens claim Irish heritage. My wife and I had many cousins scattered over many states and we knew this would be an opportunity to look some of them up.

After the merger of Oral-B and Braun, there was considerable overcapacity within the organisation. Furthermore, the use of third-party manufacturers was increasing, so competition was fierce, and any site that did not have a strong Continuous Improvement philosophy was not going to survive. The decision to close the Iowa site had already been made, so I knew that, even if we did all the right things, it still might not be enough to reverse the decision.

When I arrived, what awaited me was much worse than I had anticipated. There was a real lack of engagement from senior managers on the issues that mattered. They were simply not working on the problems that were all around them. There was no meaningful communications process that allowed employees to voice their opinions, which meant they weren't being listened to nor responded to. Town Halls were held each quarter, but they only shared the *good* news, while the problems were ignored or glossed over.

A project on site closure had been active for about two years. I inherited a huge file from my predecessor, but apart from the HR Director, *it had never been shared with the site leadership team or the workforce.* I found this to be incomprehensible. A large factory with upwards of 750 full-time people, and a part-time force of 150, were walking towards a cliff edge and they didn't know it!

What they were good at was getting product out the door, but it was at far too high a cost. They were good at meeting deadlines and, due to the commitment of the employees, quality was high. They had an impressive pick 'n pack operation that shipped 1300 parcels per day to a database in the USA and Canada, which included over 76,000 dental offices.

Oral-B had an upper hand in this case—it was founded by a dentist and enjoyed a terrific reputation within that community, with good reason! They made, and still make, the very best toothbrushes. But this relationship with the dentist community was a strategic weapon that many in the wider organisation did not appreciate or even know about, and it would become a key weapon in our arsenal to keep the site alive.

One of the largest, globally recognised dental colleges is located on the university campus in Iowa City. They have a conference hall named The Oral-B Room. Each year, dental freshers would visit the factory for a presentation and tour. These were terrific endorsements for our products. I also saw these features as weapons in our fight for survival.

However, despite the opportunities I saw when I arrived, there were some steep hurdles to climb. We had a couple of mid-level managers who refused to talk to each other and had not done so for years. This was unacceptable to me. Pulling from my lessons in the military, and years of experience in manufacturing, I could not care less if they disliked each other and refused to socialise outside of the workplace, but it was unprofessional to behave like that within the workplace. These mid-level managers were among the earliest departures from the plant.

What startled me most was the lack of urgency. In my initial walkarounds and observations, I estimated that, at any time, at least 30% of our equipment was shut down and waiting for either raw materials, technical support, spare parts, or work orders. Our supply chain was ad hoc and needed major change. Moreover, Warehousing and Distribution were separate departments and that made no sense to me. Planning, Material Control, Supplier Development, Warehousing, and Distribution are inextricably linked and perform much better when operating as one unit.

At every level, accountability was as close to zero as you could get. I found the workers on the factory floor to be decent, honest, and hardworking people, but they had been failed by poor leadership and inadequate processes.

The level of manual operations was quite astonishing. This was the most expensive site in the world of Oral-B, and we had already automated many similar operations in cheaper locations like India and Mexico. Although they had installed some robots in the Moulding Department, similar to what we had done in Ireland, months later they had still not withdrawn the people previously performing those tasks. When I queried the Production Manager, he told me the people didn't *want* to transfer. I told him, in no uncertain terms, to have them out by the following Monday or I would get someone else to do it. Eight people were transferred.

Another surprising feature was the length of time people were left doing the same jobs. Some employees had spent *40 years* doing pretty much the same tasks. No meaningful training and development plans existed. For shop floor personnel, training was confined to new equipment when it arrived, or new products. Nothing like team building had ever been carried out.

When my changes kicked in, many people were forced to move and learn new tasks. This led to some choosing to retire, while others sought out therapy, as they were traumatised and frightened by the changes that were coming. I only learned of this when I received a message from the Chief Medical Officer at Gillette HQ who monitored trends and costs in the factories. Apparently, there had been a steep increase in those seeking therapy. This made me angry, and not at the employees.

These workers had their potential stifled for decades, and some couldn't cope with even the most minor of changes. It is wrong to treat your employees this way, and it abdicates a key management responsibility:

You must offer every employee the opportunity to grow.

It was with some satisfaction that, when the project was well developed, many people from the factory floor thanked me for changing and enriching their lives. They were enjoying work much more because the changes and their increased involvement had given them an added confidence.

One thing I was apprehensive about when moving into the Iowa environment was the potential for union activity. Before I left Ireland, a union official friend had told me that I would encounter stiff resistance to radical change wherever the Teamsters Union was involved. He had met some members at various forums around the world and they were still very traditional in their thinking. The Teamsters were the union in Iowa for most of the shop floor personnel.

Along with my HR Director, I had an early meeting with the Teamsters area organiser for that region. I spelled out why I was there, what I felt was possible, and told him that a workforce of 450 in a healthy factory with a strong future was preferable to one with 750 that was doomed to close. If I were to encounter serious resistance and disruption, the site would have no chance.

In fairness to him, and to my surprise, he listened carefully to the points I was making, including plans for investing in the development of the entire workforce. I showed him some of the correspondence concerning closure that had been around for two years. We spent most of an afternoon together, and at the end of our meeting, he said he would ensure I got the space to do what I felt was necessary. He could not guarantee what some individuals would do, but they would not have his support, provided what we were implementing was done fairly.

I will always be grateful for his tacit support. Not all union officials behave with such maturity. We shook hands on it, and I must say I never had any serious difficulty from the Teamsters Union.

Among the first things I did was meet with the workforce to give them the hard news. This was difficult and many refused to believe it. If you are told for years how good you are, that you are world class and better than everyone else, it is difficult to hear a message like the one I was delivering:

"The plant *will* close if there are not major changes."

Because many did not believe me, I posted some of the communications around the factory, confirming my message. Many were hurt by this and shocked that they were only finding out about it now.

I also told them that I believed we could turn the site around, but I would need their support and participation to do so. I stressed over and over that we would seek their input, listen to it, and respond to it.

But I made no promises about survival. I made it clear that I thought we could create a great plant, but that it may not be enough to reverse the decision.

I brought over Clem, an experienced and effective consultant from Ireland that I had worked with before. I had him run workshops with all the shop floor personnel in groups of about 15. It was important that people could speak freely, and for that reason, supervisors and office staff were scheduled separately.

The teams were asked four important questions:

- What is working?
- What is not working and what are the barriers to that?
- What should we retain and do more of?
- What do we need to do to become world class?

This is a terrific way to really get a feel for how people are thinking, and I've done this at numerous locations across the world to great success. Usually, by the end of a session, the walls are plastered with all kinds of comments, but then we get them to vote on the top three. Without fail, people respond maturely to this request, and the silly or unrealistic stuff is discarded.

In this case, after narrowing it down with all the work groups, we collated all the information and posted it around the factory. Appropriate people were allocated the tasks with due dates for completion. These were updated weekly.

I brought in a terrific Lean expert, Keith, from London, and he has remained with me ever since. With his help, and the feedback we received, we created implementation teams for each of the following areas:

- Reorganise warehouse/Kanban to factory floor
- Kanban for internal movement of WIP, and introduction of Backflushing
- Creation of a Central Store for spare parts
- Reorganisation of resources to ensure maximum support for production
- 5S: A Place for Everything and Everything in its Place

- Inclusive Communications Process
- Supplier Development/Kanban to warehouse
- Fast Changeovers - SMED
- Autonomous/Preventative Maintenance
- Automation of high manual operations

It was early in 2000 when we really started the Change Program. Factory floor personnel were on every team and were enthusiastic and anxious to contribute. Never before had they been involved in anything like this. We ran regular Kaizen events and Process Mapping with terrific participation from everyone.

I recall a workshop with my leadership team around that time, and the external facilitator we were using came to me during a break and told me that some of the team did not understand what I really wanted. This puzzled me, because I believe I am a reasonable communicator, but we discussed it, and he suggested that I try to express my expectations as simply and directly as I could.

In 30 minutes, I wrote the following Ten Management Operating Principles, which represents what I believed then and now. You may notice that many of these Ten Principles were derived from the lessons I learned over the course of my career.

Ten Management Operating Principles

1. Open, honest environment where good communications are treated as an essential part of our process and in which everyone has a genuine voice.

2. High awareness of the competitive environment in which we operate and a constant focus on cost reduction and continuous improvement.

3. Urgency and speed of response is built into all of our processes through the empowerment of people and elimination of bureaucracy.

4. Where people are acknowledged as our greatest asset, and their conditions of employment and personal development are invested in on an ongoing basis.

5. Where goals and objectives are clearly communicated to every level, and each person is held accountable for those.

6. An organisation that is in constant touch with developments in technology, the markets, and the wider organisation, and anticipates and prepares for change rather than being forced to react to it.

7. Where financial systems are transparently designed so that clinical measurements of each of our key areas of cost are available as tools for managers and team leaders to effectively manage their areas and be held accountable for them.

8. An effective appraisal system where all employees will be given an assessment of their strengths and weaknesses, with follow-up plans to address those areas that need improvement.

9. Where swift action will be taken to deal with personnel who obstruct or do not sufficiently contribute to change that is necessary for the ongoing development of our business.

10. A dynamic, supportive, empowered team-based matrix organisation that allows for fast decision making and where occasional mistakes are tolerated and used as opportunities for learning.

The management principles are deliberately direct, and the language is unambiguous. I felt they had to be this way to get their attention.

We threw away the previous "Site Vision" posted in reception. It was full of flowery language, saying the usual stuff about valuing people and community. It was obvious to me that this wasn't the case in years previous. Leaving people in the same jobs for decades while neglecting to provide robust and effective processes so they could work effectively was *not* valuing people.

We held workshops for the entire work force to present and discuss the Ten Management Operating Principles to ensure everyone knew what they meant. We posted them in reception and around the factory. We audited sections of the workforce over the following 18 months to hear how they thought we were performing against them. When I was happy they were sufficiently embedded, this was no longer necessary.

I also have a number of slogans that I have erected around the factories I've managed over the years.

For example, "Nobody Waits" is meant to make employees think about what they are waiting for, and whether they are keeping others waiting! Due to the seriousness of the situation at the Iowa plant, I turned this into one of our Tools. Since then, I have been using it successfully around the world. At our Town

Halls and other forums, I challenged everyone at every level to ask themselves why they waited. I told them they were entitled to express their frustration and to make it clear that it was unacceptable. It worked a treat and was key to injecting the kind of urgency we needed to make our program successful.

Plant slogans used in Iowa

Three months into the transformation, I began to separate those I knew who would never contribute to the changes we needed in the organisation. A couple of senior managers moved on, followed by several more at middle level. Later, a number on the frontline would also move on, as well as others in some key positions. I saw all these as necessary and not as a negative reflection of the Change I was administering. There *will* be people moving on when you institute Lean for the first time. This is natural and even welcomed.

I created one supply chain organisation by combining Warehouse, Pick n' Pack, and Distribution with Materials/ Planning. The Materials Head soon departed, so I promoted the Distribution and Warehousing leader to head up the combined departments. Bill was an inspired choice and did a terrific job supporting the new practices and producing great results.

An experienced HR director was heading up Manufacturing, but after one long factory tour, we both agreed he should revert to HR. The Quality Assurance Head had just departed, and I saw an opportunity to cut overhead, so I combined this with HR. The guy heading up HR, Jim, was an outstanding young manager with a sound engineering and manufacturing background, so I had him take over Manufacturing. Jim remained there for a year and did a great job embedding all the mandated changes. He then took over Engineering, leading a major reorganisation and a sharp increase in automation. Today, he is now a member of our Altix consulting team.

Mark, a young Irish guy that originated from the Newbridge site, had previously been assigned to Iowa to develop a new product and bring it back to Newbridge. He was so impressive that he was rapidly promoted, and after the project was complete, he was transferred to HQ in Boston to work on Program Management. But I wanted him back in Iowa, so I convinced him to return and head up Manufacturing. Mark was a major contributor to the successful transformation in Iowa and went on to have a stellar career running engineering improvement projects as a Director for Gillette and, later, Procter & Gamble.

My team was in good shape!

Another major change I insisted upon was that leadership should be far more visible to the workforce. So, I ensured that both I and the leadership team spent time on the factory floor engaging with people. I call this **_formal informality_**. This way leadership stays in touch, we acknowledge strong performance, listen to their issues, show genuine interest, and we are visible to the greater workforce.

I see this as a critical part of leadership—something that I initially witnessed from Kevin, our leader at Oral-B Newbridge. No one was exempt. I applied this as a rule of thumb for each position. Obviously, I did not expect to see the Finance Head there as often as one of the operational guys, but I did expect to see him, and I expected everyone to be able to contribute to solving issues on the shop floor and be ready to discuss them at our weekly management reviews.

We also did many things to change the image of the site. I used every trick in the book to get as many influential people as possible to visit from Gillette and Braun HQ. We ensured that a well-thought-out program was developed for their visits, and our goal was to give them the best possible memories of their time at the factory and Iowa City, from the moment they got off the flight until we dropped them back at the airport. We tried to find out what kind of food they enjoyed, and whether they liked to have early nights in the hotel or preferred to visit the bars. We had a terrific Irish bar there called The Dublin Underground that was popular with many. We also ensured that, as their largest customer, the local Sheraton Hotel took special care of our guests.

At the factory, no stone was left unturned. We always strived to make the tours as interesting as possible, and we involved people at all levels. This was also a great experience for our factory personnel, as previously they would have had no such exposure to VIPs.

We created a plant slogan: *"Fighting for Our Future."* We emblazoned that on an orange shirt and gave every employee two of them. When visitors arrived, they were met by an impressive sea of orange. As I bid a Vice President from HQ goodbye, I recall him saying to me, "Jeez, Cassidy. You guys mean business!" By Fall of 2000, we had great excitement in the factory and results were improving visibly on all KPIs.

We were on our way!

Oral-B Iowa: Fighting for our Future!

As we moved forward, even more opportunities arose for improvement. Each area was responsible for ordering and storing its own spare part requirements. There was no organised system to manage them, and we were frequently running out of items. Many of these came from Germany, so lead times were lengthy, and this could not continue. We created one central stores, put someone in charge of it, and created min/max levels just like our Kanban in production. Within three months, shortages were a thing of the past.

The factory had a bewildering number of job classifications which caused all sorts of confusion and demarcation lines. I was used to working with about 10 classifications, but Iowa had 33! This had to change, so we negotiated with the union to drop it from 33 to 20 in Year One, and the following year we brought it down to just 9. This was still about three too many for my preference, but we don't always get everything we desire. We'd made great progress, and it had been a significant concession by the union representatives.

An interesting feature of the Oral-B Iowa workforce was that many came from farming backgrounds. They were hardworking and used to working with their hands.

Show me the farmer that will call a plumber for a simple leak or broken fan belt!

I quickly found that, as we implemented good processes, our floor staff were able to fix many of the simple things that previous demarcation would have prevented them from doing. I often say the Iowa team were the most skilled "unskilled" workforce I ever worked with. By "unskilled" I mean they had not completed an apprenticeship.

Likewise, the engineers, technicians, and toolmakers were terrific and innovative. They were full of ideas to implement in the new environment we had created, and rose to the challenges brilliantly.

Within two years, productivity was soaring, and as I had predicted, we had to deal with the surplus people our changes and automation was creating. So, we obtained agreement from HQ to offer separation packages to outgoing staff.

As many employees had worked there since the factory's foundation in 1958, the generous separation packages were attractive to many, and by the end of Year Three we had reduced our workforce from 750 to 450. Additionally, we had a fluctuating temporary workforce of 50 to 150 personnel building product promotional stands for our larger customers like Sam's and Walmart. This type of work was ideally suited for outsourcing, and this is exactly what we did.

Many factories struggle when they have a large array of processes. It always makes sense to choose what is most important and what will best "fit" your space, equipment, costs, and environment. Having a sensible outsourcing strategy is key to having an ordered and effective operation.

However, despite all our efforts, I was still getting sceptical feedback about our chances for survival. I have found that, once the word goes out in large corporations that a particular site is doomed, it is very difficult to reverse that thinking. So, we needed something else.

From our workforce representatives we asked for and obtained a pay freeze for two years. I knew this would get special attention and it worked. After this move, some of the senior people in Germany and Boston were finally accepting invitations to the plant. It was the final push we needed to change the organisational tone around Iowa.

It was in November of 2003 that the President of Braun Oral Care visited our plant and declared that our site was safe and investments would continue. It was terrific news and a great relief to our entire workforce. They had listened to the harsh messages I had delivered, then responded magnificently and implemented all the changes with great results. We'd saved hundreds of jobs, and it was a good thing too; the US, and much of Europe at that time, were losing factories to Mexico and China. Iowa and the Midwest had been hit particularly hard, and many small towns were surrounded by derelict factories. If ours had closed, it would have been difficult for many to get a job with similar benefits.

But what we had shown in Iowa was that factories in high-cost locations *could* survive by doing the right things. I have been preaching that message ever since, as well as implementing similar successful programs all over the world.

Iowa remains one of my proudest achievements, and I will forever recall with fondness the memories and the friends I still have to this day.

Shortly before I moved on in the Fall of 2004, I was contacted by the Lean Enterprise Institute (LEI). They were and remain the premier organisation representing Lean Thinking and Practices. They heard from someone in Gillette that something special had occurred at our Iowa factory. Could they come and see?

So, their Communications Director, Chet Marchwinski, came for a quick, two-day visit to determine if this was real or not. He decided it was and came back later for a week to complete his work. I am forever grateful to him for his patience and his deep understanding of what had been accomplished.

The result was a written Lean Case History that documented in great detail the changes we implemented in Iowa. Since then, it has been translated into at least six languages, including Mandarin and Korean. It can still be found in their online library, called 'Toothbrush Plant Reverses Decay.' It can also be found on our website at:

www.altixconsulting.com.

It was a fitting finale to one of the most satisfying assignments of my career.

Lessons learned at Oral-B Iowa City that I carried with me for the rest of my life:

- ➢ Not all union leaders have closed minds
- ➢ Entire workforces can believe they are the best when they are clearly not
- ➢ When done right, what you can accomplish with Lean practices is limitless
- ➢ Factories can survive in high-cost environments by doing the right things
- ➢ Strive for consensus, but when you don't get it, use your instincts and experience to make a decision. The workplace is not a democracy

CHAPTER SEVEN

Iowa to Ireland

Braun Oral-B Carlow

In 2004, I returned from Iowa to run the Ireland–India Campus. By this stage, the campus was well established: Oral-B Newbridge, Braun Carlow, and Oral-B India. Three plants in all.

I was given indications from my superiors that it was unlikely the Braun plant in Carlow would survive. It had a long history of poor industrial relations, reminiscent of the usual strikes and unrest I had encountered at General Motors. Even though they were part of Gillette, Braun had been left in a state of semi-independence, because, at that time, Braun was still very much run as a German company, with little movement of people between divisions.

There was a strong hierarchy, remote from the real issues, with downward communications only.

Are you starting to see a trend?

However, Braun Carlow also had terrific strengths. Braun is a very strong engineering-focused organisation and they are highly innovative. In the Carlow factory, there were layers of technical strengths in engineering as well as tooling and crafts personnel. Like all Braun sites, it was large, well laid out, and maintained to the highest standards. It had multiple technology platforms, so it was well worth salvaging. Another remarkable feature was that, although 70% of the factory floor personnel were female, not one was working at a higher grade. *This was in 2004.*

I did what I had done in Iowa and Newbridge and brought the same consultants, Clem and Keith, with me. Both were seasoned and experienced. They knew how to change how workforces think, and I was confident they could lead the implementation of new work practices. For the first year, I also brought in Frank from Packard who was sorely needed for his experience working in facilities, manufacturing, and HR. He was a veteran manager, good with people, and I needed someone to quickly embed the changes on the factory floor. Moreover, I would be traveling regularly, so I needed eyes and ears I could trust and someone to help decide who was contributing and who was not.

First, we had to confront the unions. These were dominated by members of the craft group who had become arrogant and were used to getting their own way. They had all kinds of negative demarcation practices and were largely a law unto themselves.

At our first Town Hall, I bluntly told the workforce that I was sent here with a mission: either to salvage the plant or close it. I said that I believed if we implemented a Lean Program of Change that we had a good chance of survival, but if it was resisted and the workforce continued to be controlled by a tiny number of disgruntled union activists, we had no chance. I told them that they would be involved in helping design the program and its implementation.

Furthermore, I appealed to the 70% of the workforce that were female. I asked them to consider why not even one of them worked in higher grades! Did they think that was fair? I asked them to look at those who were preaching non-cooperation. They always sat together, they were all male, and they were all earning considerably more than they were. How did they rate the services and support these union members were providing to them? I knew it was poor.

Waiting for support was not as endemic as it was in Iowa, but it was a significant factor. I told them they were entitled to expect a world class service from anyone who was supporting production. Anyone not building product was directly, or indirectly, part of our overhead. Regardless of their level in the organisation, we could not afford to carry anyone who was not making a significant contribution.

I repeated that they could choose to follow these men in the union or listen to what I was offering, and after a time, I could see that the message was getting through. I spent hours each day walking the factory and repeating those messages.

We completed the same exercise we carried out with the employees in Iowa:

- What is working?
- What is not working and what are the barriers to that?
- What should we retain and do more of?
- What do we need to do to become world class?

For the first time, we compiled the feedback into a booklet instead of just posters, and they once again contained responsibilities and completion dates. I knew that some shop stewards were actively campaigning against cooperation with the program, and we wanted to show the workforce that we were deadly serious. We provided every employee with a booklet, and schedules were posted around the factory and updated weekly.

Soon after, six cross-functional teams were put together and the transformation began. We were on our way!

Visual Management Braun Oral-B Carlow

Meanwhile, I removed several of the leadership team from the organisation, all of which I knew were not going to be supportive of the transformation. Workforces always know who the strong contributors and who the passengers are. Seeing them moving on always adds credibility to the program.

The program picked up pace and, in a matter of months, the shop floor was transformed. The factory had an automatic, high-level storage warehouse, and its location made it time consuming to put away and retrieve product. But inventories plummeted so much that we were able to stop using it altogether, and we began relying on Kanban deliveries from our suppliers instead. Kanban is a wonderful tool to free up space, expose slow and non-moving stocks, and put the people closest to the real work in control of when inventories need to be replenished.

View of Factory Floor Braun Oral-B Carlow

One year into our program, it was announced that Procter & Gamble (P&G) and Gillette were to merge. This was a massive shock, and one of the most significant mergers of large organisations in years. Gillette was a 104-year-old, Boston-based company, proud and steeped in the history of that city. It had been acquiring other businesses over many years, and its divisions included Duracell, Oral-B, Braun, Waterman, and Parker Pens. However, several years before the merger, it had taken its eye off the ball and made itself vulnerable. After delivering double-digit growth for ten consecutive years, it had floundered and struggled. Major shareholders were restless.

In response, Gillette appointed its first ever external President, and he turned out to be ruthless but very effective. He issued a couple of diktats that were to be applied globally without exception—a great lesson for anyone who needs rapid improvements. It was blunt but it was very successful: Each division would deliver 6% profit annually net of inflation and do so with zero overhead growth.

I recall at a large meeting in Boston, with leaders from around the world, the President of Gillette said a couple of profound things. He was explaining the changes he was implementing and what he expected, when he said, "The boat is leaving the dock. You are either on it or you are not." He understood many within Gillette were unhappy about the changes he was spearheading, and he provided them with an ultimatum. He also said something that has stayed with me, and I have frequently used it since.

If you ask someone why they came to work today and they cannot explain it clearly in three minutes, you have a problem.

With his diktats in place, Gillette was quickly restored to good health and Wall Street was happy again. Then, he turned around and sold it to Procter & Gamble!

Although the language was carefully crafted that this was a merger of equals, few bought that line. In every such coming together there is a dominant partner. The critical decisions will be driven by them, and their culture will prevail. Such was the way with this one.

Over the following few years, there was a huge exodus of senior people from Gillette and its divisions. P&G are a terrific organisation, several times larger than Gillette, so the results of how things progressed was no surprise. Their policy was to recruit the vast majority of their people from college so the only culture they knew was P&G.

Personally, I found it overcontrolling, surprisingly hierarchical, and unnecessarily bureaucratic. Decision-making took too long, and it was often difficult to know who was actually making the decisions. They were also centralising many of the functions within their factories. Most disciplines reported to HQ, so plant managers did not have the scope that we had in Gillette. In my experience, centralisation takes away autonomy and slows down decision-making.

For someone who had embraced Lean Culture, this was the opposite. They had too many people who created a lot of non-value-added work. I learned a long time ago that I would rather have too few people in my operation than too many. Having too many causes all kinds of problems: resentment, jealousies, increased absences, and you lose that vital ingredient for any kind of business to be successful: a sense of urgency.

Interestingly, when the retired CEO & Executive Chairman, AG Lafley, was recalled to the helm in 2013, he reiterated my concern, saying that, due to his experiences on the boards of other organisations, he realised P&G had too many people, and he embarked on an aggressive reduction plan.

Despite not being Lean in their operation, P&G was and still is an incredible company. They have brilliant products, many billion-dollar brands, and terrific career pathing development for their employees. Remuneration and conditions were excellent, and they continue to be a great employer. Indeed, I have encouraged many young people to join them.

They also had their own system called Integrated Work System (IWS). This had been around for over a decade and was being applied in various parts of the world, but surprisingly was not practiced in any of the USA factories at that time. There was an immediate demand that all the Gillette factories and their divisions commence implementation.

Personally, I and other Lean thinkers did not like it. It had 11 Pillars representing what they considered to be major processes, which was fine. It also had many attractive features, particularly in TPM, but I found it overly bureaucratic, and the more I looked at it, I saw that, in order to implement it, we would need resources in excess of what I was accustomed to having in my factories. I felt that what I was doing up until then would get results much faster and with fewer resources. Moreover, it only applied to factories, which made no sense to me.

In every factory where I have implemented Lean programs, many problems had their roots in HQ, and the issues often had to be dealt with there first. If not, significant changes needed for the good health of the factory will go unchanged. Most often, these changes are how planning and forecasting information is provided and how new products are introduced. There are other issues, but these are the big two in my experience.

I knew that this merger would change things. Now we had three plants in Ireland. The one we gained had a lousy reputation for poor industrial relations, and it was not highly regarded within P&G.

Regardless of the earthquake this caused within our world, we continued with our program in Carlow, and made great progress with results improving on every front. Up until now, I was confident that the site was looking increasingly attractive for future investment, and I was hearing the same from senior people in Braun and Gillette.

But as the weeks and months passed, my confidence waned. When organisations come together there are always consequences. Overnight, we had gone from about 37 factories worldwide to about 150. Some of these were not going to survive. Synergies would be sought, and many closures were inevitable no matter what we did.

I heard rumours that they were considering closing my old Newbridge factory and moving it to Germany and Poland. This plant was still in good shape, but things had stagnated. In response, I made some changes to the leadership team, and we were back on track.

I contacted the Head of Oral Care, telling her I had heard the rumours of plant closure, and invited her to see both our plants in Ireland. I told her I believed her P&G factories would have to be good, *very* good, to be better than ours, and to please not to compare us to their own factory in Ireland, as we were streets ahead of them. She came and liked what she saw. The rumours stopped, and we got a lot of support from her after that.

In the end, the program was outstandingly successful in Carlow and very similar to what we had done in both Iowa and Newbridge. We had carried out the employee survey with the four key questions. We converted that into a booklet and gave every employee a copy. We made schedules to fix all the issues we agreed upon, then nominated who was responsible to deliver and when. We posted them plant wide, updating them weekly. We had six teams implement the major changes in each key area:

- Communications
- Fair & Equitable Treatment for all

- Total Productive Maintenance (TPM)
- Total Quality Management (TQM)
- Logistics & Supply Chain
- Training & Development Plans for all

In my experience, these are the key areas that are commonly addressed in factories I consult with today, with the outlier here being "Fair & Equitable Treatment for All." In Carlow, there was a lot of mistrust, no women were working in higher grades, and some supervisors and managers were being accused of favouritism. It was imperative that it should be included.

When our program was complete, the key results were:

- Support headcount to production reduced by over 50%
- Total inventories reduced by 50%
- Inventory turns increased from 6 to 20
- Total savings over 2 years $12mm equal to over 16% of total manufacturing costs (TMC)
- Full Kanban deliveries of raw materials from warehouse to factory floor
- Kanban deliveries on regular raw materials to factory
- Work in Progress transactions eliminated by introduction of Backflushing

The atmosphere in the plant was transformed. Employees were engaged and welcoming change. The negativity from most union members went away. Some women were now working in higher grade jobs, and everyone was participating in various activities within our Learning & Development Plan.

Every two years, P&G would celebrate a small number of significant achievements that occurred around the world, and in 2007, I was invited to showcase Carlow at a major event of about 200 senior people at HQ in Cincinnati. They helped us make an 11-minute video summarising our plant's trans-formation. This can still be found online and on our website.

My job in Carlow was done. Sadly, however, the merger of P&G and Gillette took its toll. A few years after I left, the Carlow plant closed, but not all was lost. Two key products were transferred to the Newbridge site along with 150 jobs.

I always say that factories can survive and thrive if they do, and keep doing, the right things. However, occasionally events occur that supersede their best efforts.

Lessons learned at Braun Oral-B Carlow that I carried with me for the rest of my life:

- ➤ Organisations don't merge; the largest will be the dominant partner
- ➤ When your organisation is acquired, your best efforts, no matter how successful, may not be enough
- ➤ Automatic warehouse systems are not the solution for every factory
- ➤ Aggressive union activists can be neutralised by involving the entire workforce in a progressive Change Program

CHAPTER EIGHT

Ireland to Shanghai

Braun Oral-B Shanghai

In 2007, I went with P&G to the Braun Oral-B Factory in Shanghai on what would turn out to be my last assignment with the company.

Situated in one of the largest industrial zones in China, it was a large sprawling site consisting of several buildings on three floors—not the way in which you would design a modern factory, but it had started off as a tiny operation and had grown steadily over the years. Its primary products were power tooth-brushes and male and female electric shavers. The workforce was about 1100 strong, and about 20% of those were temporary.

Organisations are often criticised for having temporary employees, but in manufacturing, demand will ebb and flow, and temps protect the permanent employees from layoffs. My message to temps was always, "If you show the kind of attributes that we are looking for, after some time you will graduate to a full-time position, and you will then enjoy the security that comes with this approach." It also suits many who do not wish to commit to full-time working.

In the West, China is poorly understood. Memories of communism as we knew it during the pre- and post-WWII era still dominates much of the media's comments and the world's thinking. Nothing could be further from the reality.

Yes, it is a centrally controlled, one-party state, but its population of over a billion have been catapulted out of poverty in a few decades. You do not find people living on the streets like you would in major cities all over the world, including the advanced Western ones. Beggars are also rare compared with the major Western cities.

Contrast this with neighbouring India. Four decades ago, India's economy was far ahead of China. Today, China is light years ahead on every front. India's infrastructure lags far behind China's, and many millions of its people live on their streets. The disparity in wealth between India's rich, middle class, and poor are staggering. Not so in China where the whole population enjoys a higher standard of living.

I am convinced that, were China a Western-style democracy, they would not have achieved what they have done. India inherited a democratic system from the British in 1947, long before it was ready for it, while China concentrates on the "greater good." Yes, this can be at the expense of some individual freedoms, but this pales into comparison when one looks at what it has achieved. Its population understands this, appreciates it, and is puzzled by the Western fixation with it.

One significant project completed during my time there epitomises this attitude. The fast train connecting Beijing and Shanghai, with a track 1320 kilometres long, was a three-year project completed on time and within budget. It shrunk the travel time between China's two major cities from ten hours to four. With around 300 million people using this service annually, this was a huge gain to the economy.

The project team spent two years studying railway systems around the world. Obviously, huge disruption was going to occur as building progressed, particularly in urban areas. In the West, negotiations with vested interests would take years and cost billions. However, in China, they had simple, common-sense solutions applied at every point, without exception.

First, they selected three potential routes and gave them to the local party branches so they could meet with local groups and debate and argue for and against. They would consider the feedback, but regardless they were going to select one of them. Then, those who would have to move house would be given a list of locations and they would select three potential homes in order of preference. They would not be guaranteed their first or second choices, but they would get one of them.

One of our employees was impacted by this. Initially, she was not happy and attended several meetings where she and her neighbours voiced their opinions. However, where she lived was on the route chosen, so she had to move. She was philosophical about it, felt that those impacted had been listened to, and saw that it was for the greater good. In developed democracies in the West, such thinking no longer applies, and large projects such as this one are almost impossible to contemplate.

Even with this difference in attitude and culture, I anticipated the issues I would encounter in Shanghai would be similar to those elsewhere, and so it transpired. Still, I was aware that there were strong cultural differences between us, and I set about understanding those as much as possible.

I was happy to see some Lean practices had already commenced before I arrived. However, they were only being used within Production and weren't part of a site-wide plan.

Implementing a Lean process in isolation like this has very limited benefits and will not sustain itself long term. Only a site-wide plan that includes suppliers (and HQ when necessary) will work. In this case, I quickly discovered, as I have in so many other sites, that the Materials & Planning systems and processes were poor. If you have poor systems, it doesn't matter what you do elsewhere; your improvements will be hampered and undermined.

Someone once described the Planning & Materials system in a factory as akin to a high-rise apartment with all the electrical wiring and plumbing running through it. There are so many that touch each other and intersect that to look at a drawing can be quite bewildering, except to the most trained eye. Similarly, to map a system of planning, and to understand the flow of materials with inventory levels, lead times, and off-sets, is just as complicated. It is the most common area of difficulty in factories globally, and the most underestimated and misunderstood. Regular shortages of raw materials and poor planning result in shortages of finished product, which upsets customers and threatens the business. When this happens, many people within the factory must deal with the fallout. Shortages increase phone calls, emails, meetings, changeovers, and all of those involve many people, which creates a highly stressed environment. A good system will eliminate that and bring calm to your entire operation. In Shanghai, fixing this poor system became a priority for me.

That wasn't my only hurdle to overcome, however. P&G had aggressive Environmental Health & Safety (EHS) targets, and we were a good way off the minimum expected. They had given all the plants they acquired 18 months to meet their targets, otherwise future investments may be put on hold.

Modern EHS procedures are demanding, and when being applied in an environment like Shanghai, where you often see mam, dad, and baby on the same scooter in chaotic traffic, it takes a while for them to accept that, when they are coming down the stairs, they must hold a handrail! This is just a small example of many that we had to overcome.

To make things worse, many of the facilities were in a poor state. The toilets available to Production staff were unacceptable and far below the levels expected in a global company. Cloakrooms, where employees changed their clothing and kept their personal effects, were also neglected, and many people did not have their own lockers. Some lockers had been broken.

I quickly realised that I had some strong people on my team. Chen, the Head of Production, was terrific, and in fact, years later, went on to lead the China Festo organisation—one of my largest clients when I entered the consultancy world. Lily, the Finance Director, turned out to be one of the best I have ever worked with, and she became a trusted advisor on many topics. Rex was our IWS manager, and I also gave him a site-wide role in coordinating and driving cost savings. He was ideal for the position, because he was persistent and would not take no for an answer. The Quality Assurance and Engineering depart-ments were managed by two German expats who were effective performers and helped keep communications straight with HQ in Germany. We had several strong middle management people in a number of departments, particularly in Production, and I was satisfied that we could build on that. But first, we had to sort out our Materials & Planning processes.

We also needed someone effective to head up our EHS unit to ensure it had a strong voice and would embed the new safety processes into our workforce. When I arrived, discipline was poor. Accidents or "near misses" were not being investigated properly, and Corrective Actions (CAs) were either poor or non-existent. Dangerous work practices like climbing on workstations, putting hands where they should not go, and smoking in toilet cubicles was commonplace. We needed someone capable of turning this ship around.

We interviewed several candidates through a local agency and found Chris. She had plenty of experience with a UK-owned multinational who had excellent EHS standards, and to this day, she remains the only person I have promoted during an interview. Initially, we were offering a position on senior staff, but I realised she was such a strong candidate that offering her anything less than a management position would not be attractive enough for her. Having her on the team was vital, for there were significant new projects in the pipeline, and we could not put those at risk by failing to meet P&G's EHS targets.

Upon her arrival, we arranged to meet the entire workforce in groups of about twenty. I addressed them with a key message, then I introduced Chris. She described what her role was and asked that they give her their full support. Chris went on to develop a comprehensive program that she implemented immediately. I had her report progress at our weekly management review, and this helped ensure that she had the support of the entire leadership team.

Chris's personality was ideally suited to convincing people at all levels to change habits. Our EHS targets were comfortably achieved, and since then, she has gone on to head up a number of her own organisations that operate all over China.

I had about an hour's journey to work each morning from city central to the suburb of Minhang. George, my driver, would pick me up at 6.45am, which gave me plenty of time to contemplate and plan my day. For a time, I took a Mandarin class each morning from a teacher recommended to me by another expat. We would pick her up early on the journey, and she would teach me on my way to work, and then my driver would drop her home later. It helped enormously and allowed me to communicate, at least to some extent, with the locals. I have found that, wherever you are in the world, making an attempt to communicate in the local language is always appreciated. It also led to a significant improvement in George's English!

When I arrived to work each day in those first weeks and months, I did not know what materials we were going to run out of at the factory. The Materials Manager, who had been in place for about a year, did not know either and argued that "This was normal in China."

But it was not normal for me, and I could never accept it. I sent him to our best sites in Europe so he could see what I was trying to convey to him. When he returned three weeks later, he said that he'd enjoyed it, and that he was very impressed by the materials organisations he saw there. However, he persisted in saying, "It would not work in China."

After that discussion, I went straight to HR to set about recruiting a replacement. I insisted on speaking to the best-known agency in Shanghai, where all the top candidates go. This caused some initial discussion, because the Shanghai factory had used much cheaper recruiting options in the past. I told them to discard those and only deal with the best.

It was a lesson I'd learned a long time ago when recruiting. When you set your expectations high, it will cost a bit more. But in the long run, it's worth it.

I talked to the leading agency in Shanghai and described the kind of person we would need. I told them the most suitable candidates would be found in the auto or electronics industries. This is where you find Materials Management/Supply Chains at their best.

He presented me with three terrific candidates. One of them, Xiaoying Xu (Xxy), was from a division of GM. It turned out we both spent time on a site in Portugal some years previously, albeit in different decades. He was a fantastic appointment and brought all the attributes I was seeking, including a deep knowledge of how Materials Organisations, including Planning and Supply Chain activities, should be optimised. He was patient, an excellent teacher, focused, determined, and had a full understanding of Lean and its Toolbox. I had found the key to solving the problems that bedevilled our operations.

Xxy, Supply Chain Leader Braun Oral-B, Shanghai, supporting 5S

Xiaoying Xu changed our warehouse into one with Super-market Principles, similar to what I had done in Packard Electric some years previously. He introduced Kanban deliveries from key suppliers and to the factory floors. Inventories shrank and shortages became a thing of the past. His impact was immediate.

Kanban Area – Factory in China

Later, Xxy would join me for a time in the consultancy world and helped us transform the results of a major global organisation. Today, he works for a major auto player and is responsible for a number of factories and their supply chains.

In all my assignments, I talk about the pursuit of excellence in everything we do. In the early days at Shanghai, I found it difficult to convey to my team the levels of excellence I was seeking. Our understanding was at different levels. So, several months into my assignment, I decided to bring my team, plus next-level management and other key positions, to a weekend workshop in a beautiful, small city about two hours from Shanghai. We stayed in a 5-star hotel.

To those readers who have been to China's 5-star hotels, you know that they make our 5-stars look like 2-stars. They are luxurious at every level. Accommodation, restaurants, attention to detail, and service are superb, and I wanted my people to experience this first hand. More importantly, I wanted them to think about how we might aim for such standards back at the plant. Of course, we were not going to turn our factory into a 5-star hotel, but we could aim to make it the very best we could.

Author and Team at "Excellence" Weekend

We arrived on Thursday afternoon and departed on Sunday. We had a number of themes and held workshops for each of them.

We asked: What should "excellence" look like in the following areas?

- Quality
- Customer Service (internally and externally)
- Productivity
- Environmental Health & Safety (EHS)
- 5S: A Place for Everything & Everything in its Place
- Communications
- Visual Management

I had found a New York-born facilitator who had lived and worked in China for 12 years and had an excellent command of the language and culture. He was also a fun guy, and those who have worked with the Chinese will know they like having fun in the workplace and love playing games when finding solutions.

It was an outstanding success, and when we went back to work on Monday with our Lean Program (underlined by *Excellence* in everything we do), it took off like a fire.

Apart from the reasons above, it was such a great success because the employees felt truly valued. It is often difficult for Chinese nationals to feel comfortable with Westerners. We manage meetings differently. We tend to talk a lot and try and convince others of our point of view.

Their way is different. They don't participate so much in our kind of debates, and a lot of their arguments are made quietly and away from meeting rooms. We understood that and ensured the workshops were run in a way in which they felt comfortable.

Our Mission Statement Braun Oral-B Shanghai

144

So, our programs were up and running, all our key results were improving, but we had other things to do. Our offices and factory needed upgrading. We made a major investment to improve them, but first, I wanted to communicate directly with the workforce.

I had been advised against this by the local HR leader who said that a Chinese workforce would not directly communicate with me. I ignored that advice, but it proved to be true. When I asked for questions or concerns, they didn't respond.

I wasn't happy with this, so I had question boxes installed and invited comments or questions ahead of our meetings. Before our next meeting, we received 312 questions and comments! We had these collated, and similar to Carlow and Iowa, we responded to them by naming who was responsible for solving each issue with completion dates.

Many were about the quality of the food, the cloakrooms, bathrooms, and old buses (in China, workforces are transported by large employers to and from work). There was a terrific response from the general workforce to this, and I could see that they thought this foreigner was genuine in making them a part of the Change Program.

We quickly upgraded our food menu and fleet of buses. Now they would have a modern and healthy menu to choose from each day and could travel in comfort to and from work—warm in the winter and cool in the summer.

I wanted both offices and factory improvements to take place in parallel. We hired an experienced construction engineer to oversee the projects, and then I made the great decision to appoint Ana, my admin, to oversee the office refurbishment. Later, she would move to HR, and today she is the Director of HR for a German multinational with sites all over Asia. But she wasn't the only admin in P&G Shanghai to rise to greater things.

Ana's predecessor, Doris, was equally fantastic, and I encouraged her to expand her career as well. So, she quickly moved to HR, and today she is a successful purchasing manager.

For a long time, I have believed that admins, still called secretaries in some parts of the world, are too often pigeonholed, and spend their working lives in such roles. I believe in giving them, and all employees for that matter, an opportunity to grow. All they need is opportunity and support.

With Ana overseeing the office refurbishment, I knew we were in good hands. Moreover, the Chinese are wonderfully talented and artistic, and although we had to retain the Braun Oral-B primary colours, I wanted to challenge those standards as much as we could and add a Chinese dimension. When Ana was finished, we had offices that were trendy and both local and Western in their design and colours. Our staff loved them.

Meanwhile, the factory renovations were also going great. My mantra was that nobody should have to pass by a bathroom to get to a better one, no matter where they worked on the site.

The cloakrooms, previously in poor condition, were also upgraded to our new standard of excellence. Modern lockers were provided to everyone so that their personal belongings were secure.

Suddenly, we had an enthusiastic workforce. Our Lean Program had been rolled out and was making great progress. We were exceeding all our targets. We had a new warehouse based on Supermarket Principles and material shortages were a feature of the past. Our EHS programs and practices were becoming embedded. Our facilities were upgraded, and we were attracting attention around the wider organisation as a site to visit. One year on, we then rolled out Office Lean. Processes were mapped, improvements were implemented, and information boards with critical data and information were mounted in every office.

As a reward to our workforce, we arranged a family day at a local tennis stadium. Employees were invited to bring their parents or guardians along with their children. We had several thousand people show up that day with all kinds of food stalls set up—mostly Western because that was what the staff had requested. We also had a fleet of buses bring families to the factory for a quick tour and provided each with a product pack.

A remarkable feature of the Chinese is that they may appear shy and retiring in day-to-day life but put them on a stage and they excel. So, with great enthusiasm, each department put on shows that included song, dance, gymnastics, and comedy. We had a wonderful day of enjoyment. They had never experienced anything like that before, and I was humbled by how grateful employees and their parents were. It was a wonderful culmination to my assignment as it was coming close to its end.

Author with Employees at Family Day, Shanghai

Over the course of my two years there, our results were outstanding:

- Increased productivity by over 30%
- Reduced inventories by 25%
- Increased inventory turns from 7-19
- Cost Savings of 8mm USD
- Created a new Communications Process in which everyone had a voice
- Restructured our leadership team
- Introduced a new Learning & Development Program for all employees
- Prepared the foundations for development of Value Stream Organisations
- Upgraded our facilities to World Class levels

In mid-2009, just two years after I went to China, P&G had a severance offer on the table for long-serving people at my level. Integration with Gillette was progressing, and there were too many people at most levels of management. It was a generous offer and one that I could not refuse. So, I made the difficult decision to leave multinational land where I had spent so many years.

My time in manufacturing had taken me all around the world. The ups and downs, the challenges, opportunities, sleepless nights, joys, heartaches, and all the changes that occur in large organisations had shaped me. I learned so much, and I will forever be grateful for that. My achievements to date had far exceeded the ambitions I had as a young man.

But I wasn't done yet!

Lessons learned at Braun Oral-B Shanghai that I carried with me for the rest of my life:

➤ When working in a foreign country, it may be important to integrate management styles from your background *and* theirs for best results.

➤ Create a workplace that ensures all cultures feel comfortable and able to express their opinions.

➤ Don't interpret silence for agreement.

➤ Don't let age and inexperience keep you from involving young employees in major projects.

CHAPTER NINE

Leading Change through Lean: LCL Consult Ltd to Altix Consulting Inc.

When I left P&G, I wanted to spend more time with family and enjoy my hobbies. In my multinational life, sacrifices were often made at the expense of family. Many of us who have led similar lives look back and have some regrets. Lots of family and special events are missed, and they don't come around again.

Today, I advise younger managers to keep a better work-life balance. You should give enough to the workplace but not everything. Do not neglect family or healthy hobbies. Thankfully, these days there is a better appreciation amongst most employers for that. Indeed, dare I say it, perhaps too much at times.

I find it quite astonishing the expectations of many employees; none more so than our public sector, where work-life balance takes precedence over the services they are meant to provide. There are also some employers who strive to be modern and trendy, sometimes at the expense of their company. I am all for investing in employees and their futures, but one must also expect them to be accountable for what they are hired to deliver.

One of the most common complaints I hear is of individuals not pulling their weight and getting away with it. So, it is nonsense to think that granting space and flexibility to everyone will deliver the required results. Most *will* deliver, but some will simply cruise if they can get away with it. Moreover, if there is a crisis, employees *should* work whatever hours are necessary until it is overcome.

I am frequently asked the question: "How many hours per week should I work?" My response is always: "If you are in a leadership position, you work as many hours as it takes to create stable processes and a well-run operation."

In other words, you create the environment where everyone can work a regular week and live a balanced lifestyle. I have no sympathy for leaders who complain about a long work week when they are not doing enough to create an environment that will allow them the balance they crave.

Still, managing operations in large organisations is demanding, always busy, and you never know what is around the corner. No matter if you are in a recession or experiencing rapid growth, the pace of activity that you deal with is similar. Meetings, reports, deadlines, travel, and planning for visitors is never ending. It is easy to get caught up in that and it can take over your life.

I wanted this time post-P&G to be a different phase where I could spend quality time with family and do many of the things we had talked about but never got around to. However, I did not wish to stop working entirely. So, I founded my own consultancy company: LCL Consult Ltd (Leading Change through Lean) with bases in Dublin and Shanghai.

I considered joining an established consultancy company

but decided against it, because I wanted the freedom to operate as I chose. Unlike many in the consultancy world, I did not wish to spend time with clients who were not ready for the radical changes I would be bringing. I had seen too much failure in manufacturing throughout my life, and I knew what worked and what didn't.

My office in Shanghai

Successful change takes total commitment from the most senior people, and they must be relentless and ruthless, when necessary, to ensure a successful factory transformation. Far too many people compromise on this and that is why the failure rate is so high.

Many consultants hesitate to insist on what is required, either through ignorance or because they will not put the project (and revenue flow) at risk by demanding that leadership does what needs to be done. In these cases, we see short term improvements, but they are never sustained.

With LCL, I was clear with potential clients on what was expected. We would offer them a pathway that would bring results far beyond their expectations, but they had to follow that path. This often included the removal of senior managers that, for whatever reason, were unable to stay on course.

In my experience, some traditional command and control managers find it impossible to change and will disrupt and seriously slow down the implementation. It is imperative that they be separated from the company for the greater good. We would work hard with them, both collectively and individually, to convince them of the necessary changes, while attempting to understand their fears and insecurities. But when that didn't work, they would be shown the door. Many felt they would lose control by delegating and eliminating much of their own work, but what they failed to understand is that they were constantly spending time solving yesterday's problems.

As you now know, Lean thinking and tools will deal with that and bring calm to an operation. So, the question I often heard was:

If managers are no longer putting out fires, what becomes of their job? What do they do then?

What they should be doing in the first place! Planning ahead, looking at the opportunities and challenges coming, then preparing for them so their operations can prosper and grow. Moreover, with time away from firefighting, they can spend more time on "formal informality"—walking around daily in a relaxed way and engaging, listening, and identifying problems and opportunities with employees at all levels. As you've read, it is both a great way to stay in touch and keep employees engaged and motivated. It also adds terrific value to leadership meetings when discussing problems, because consensus for tackling issues is arrived at easier with broader understanding.

These days, I also run a short workshop for some leadership teams when I think an unhealthy gap exists between them and their employees. It's called "Stand in Their Shoes." I remind these managers they did not suddenly arrive in their positions. They climbed their way up the ranks. I ask them to remember the sleepless nights that they had on their journey; the worry and insecurities they felt when they didn't know where to turn, and what it felt like when they had a boss that didn't listen or understand. Then, I ask them to put themselves in their employee's shoes and consider how they are feeling. It is remarkable how humbling this can be, and I have seen a marked behaviour change in many leaders afterwards.

Shortly after I started LCL, I was contacted by a previous colleague (and later my boss when we were in the Braun Oral-B division). He had moved on to head up global operations for Festo, the leading German global industrial automation organisation. He asked me if I was interested in supporting some of his factories. I was delighted to accept as he knew what I could do, and I knew I would have his full support. We supported their factories and supply chains in Czech Republic, Germany, Hungary, China, and the USA. It was a most successful collaboration with terrific results. Over the years, we had many other projects in Asia, the UK, and the US, also with major multinationals.

But not all my ventures were successful. I had a client that I parted from because, no matter how hard we tried, we could not help them. However, it was not for the usual reasons. It was a most uncommon situation and one that I had never encountered before nor since.

I was asked to visit a site in Asia with a workforce in excess of 3000 people. Their big problem was lead time through their factory. I spent a week with them, concluded my survey, and made a proposal which was accepted. However, I had observed that their workforce was approximately 30% too large, which meant almost a thousand people would lose their jobs. This was a lot of people. I also told them the number was likely to increase as we got into the application of the Lean Tools.

This caused huge consternation as they did not wish to separate people. The employers were very caring and paid their people well with all kinds of benefits. The result of this was that there was no employee turnover. They treated employees better than any other employer around, so why would they leave?

After several visits and lots of debate, I told them it was impossible for us to be effective. I just didn't know how to implement Lean Tools (that would have undoubtedly shortened their lead times) while holding onto all the spare people. I suggested leaving them at home and just keeping them on the payroll, but they said no; this would demoralise them. I suggested a shorter working day, but the answer again was no; this would make them unpopular with the government and neighbours in their industrial zone. Labour rates were a tiny fraction of their costs, so this was of no concern to them.

They pleaded with us to just implement Lean and ignore the fact that there were spare people. We could not do that because the people were covering for each other, sharing the work, and pretending to be busy. The company even offered to increase our rates if we would stay and find a way, but it was against the lifetime of training that we had. All our instincts were to identify waste and eliminate it. We could not find a way around it, so we reluctantly walked away from a lucrative project. They were the most loyal and caring employers I have ever met.

While consulting with LCL at one of our largest and most successful projects at the Festo site in Jinan, China, I met their

Head of Operations, a French national named Yannick Schilly. After some early hesitation, we got to know and trust each other. We got along really well, and we helped transform the Jinan site from the poorest performer in the Festo organisation to the very best.

Later, he would move to the USA and lead the transfer of Festo's operations from New York to a green field site in greater Cincinnati, Ohio. The project included the transfer of 25% of its employees, local recruitment of the rest, at all levels, and a successful start-up to its manufacturing and distribution operations. It was a challenging project in every respect, and LCL provided support for him on that journey.

When Yannick's next assignment became due, he instead decided to realise his lifelong dream of running his own business. So, Yannick, along with another French national, founded Altix Consulting Inc. He asked me to become involved, and I was glad to add some early informal support to a good and trusted friend.

He and Anne Cappel, his co-founder, embody the same values that I hold dear. Now in its third year, and despite COVID, it is doing very well and growing steadily. With its head offices in Mason, Ohio, and its regional HQ in Strasbourg, France, they plan to open an Asian office in Shanghai over the next year. It has a growing team of seasoned partners throughout the world, all with successful careers behind them.

A year ago, they offered to buy LCL Consult, and I was happy to accept. So, now I have entered an exciting, new, and may I say, unexpected chapter in my life. I never previously envisaged joining another consultancy, but I am enjoying it and look forward to continuing to transform factories and supply chains for some time yet. Helping an operation turn around and seeing its workforce becoming enthused and energised by results that exceed expectations is still one of the best adrenalin rushes for me!

Consultancy is a much different world from the one I experienced in multinational land. Of course, I still work with factories and supply chains, but now I must work through people. This took a period of adjustment, but the challenges still give me great satisfaction.

Yannick Schilly, President & CEO. Anne Cappell, CMO.
Joint Founders Altix Consulting Inc.

Many people ask what to look for in consultants when seeking support. I always give them the same response:

Check their Case Histories!

Do not skip this point. If they are as good as they tell you they are, they should have plenty of them. Speak to the leadership at those sites. If possible, visit at least one of their previous jobs. If you do this, you will be guided to a decision. We have many case studies and would be happy to show you!

View of Production in Czech factory

Kanban Visual Display

Kanban example – Festo Czech Republic

Author with team members – Festo Jinan

CHAPTER TEN

Managing Change: Why it Works & Why it Fails

In this book I have discussed how I approach Change and how I have been successful. There are so many factors involved, and not just the Lean Toolbox.

I have outlined how the application of common sense is often better than a complicated solution. Moreover, it is crucial that you always involve employees in the development of solutions to their workplace. Develop a well-thought-out communications process that gives everyone a voice. Invest in your employees' development, and ensure they have the opportunity to grow. Make it clear what is expected from them, and then hold them accountable. As I have said over and over, the lack of accountability is one of the biggest organisational weaknesses I have seen across the world.

A paper I wrote for the magazine of the Chartered Institute of Logistics & Transport summarises much of my approach. This essay echoes what you will have read in previous chapters, but I believe it bears repeating, as the lessons here are integral to the health and long-term success of your factory. After all, that is *why* you've read and bought this book.

Managing Change: Why it Works & Why it Fails

Overview

In a career spanning over three decades working with major multinationals, change for me was never ending. Multinationals are marvellously constructed organisations where, regardless of the economic environment, pressure for results is never ending, and relentless. If you are expanding, the bar simply goes up, if you are reducing, you align your resources to match that, which means shedding resources and trimming costs. In either situation, change has to be managed and although there are challenges with both situations, they are more difficult with the latter where difficult decisions must be made, communicated and executed.

For the last 15 of those years, I have been implementing a formula that has been highly successful and provided great results. I have turned ailing and poor performing factories around, creating what is known within the manufacturing world as Benchmark sites. That is, they become a model for the rest of the organisation to imitate.

This article is therefore written from the perspective of a Factory Manager. It is not an article with academic references. Rather, it is based on my own experiences witnessing and being part of both failure and success, and gradually evolving to a formula that has provided outstanding results even in situations when others said it would not be possible. I have learned much during that time, but one thing is clear: "nice" managers are not successful. Neither are they respected. By "nice" I mean those that sidestep sensitive and difficult issues. They give similar Performance Appraisal ratings to high and poor performers alike. Sensitive issues involving people are ignored. They conform and will not challenge or "rock the boat." Managers that confront difficult issues, that challenge the status quo, which will not ignore people related issues, will be successful.

I have successfully managed factories in Ireland, USA, India, and China. What I will describe here is what I do when I go to a site, how I engage with the workforce, how I deal with people who are going to slow down what needs to be done, and the tools and techniques applied that have consistently provided spectacular results. Space does not allow me to go too deeply into detail, but I hope readers will get a good idea of what guarantees success, and what will ensure failure.

The threat is real - communicate that!

In every case where I managed factories, competition was fierce. Within multinationals each manufacturing source is constantly being compared. You fight to hold what you have and try everything at your disposal to attract new investment. It is essential that this message is imparted to the workforce and repeated at every opportunity. Sometimes the threat is very direct, like when I was sent to a factory in the USA to prepare it for closure within two years, or in Ireland immediately after my then organisation, Gillette, was acquired by Procter & Gamble, and they had tentative plans to close our plant and move the business to their own legacy site in Germany and to a new one in Eastern Europe. In such circumstances it is easier to communicate as the threat is clearer. In other circumstances the threat is more indirect. I have seen many factories enjoy good years, get complacent, and lose their business to other sites that have become more competitive. Those stories are shared. Examples of factories that either no longer exist or are losing business and therefore also losing jobs are shown. They are made aware of the on-going pricing comparisons that are always taking place, so that a constant awareness of the competitive environment is being created.

Communicate the Vision

I begin by laying out the challenges, threats, and also the opportunities if we implement necessary change. A vision is created of where we can go to. Usually this says something like: 'We will be among the best in the world in which we compete' (In reality I mean "The Best", but to declare so will create too much resentment from sister factories and the senior people associated with them). I tell them that I have a toolkit, which I can apply with their support, but before completing the formula I need their input. So, I ask them 4 key questions:

- What is working?
- What is not working and what are the barriers to that?
- What should we retain and do more of?
- What do we need to do to become world class?

External facilitators are used to conduct this exercise with the entire workforce in groups no larger than 20, so that involvement and confidentiality is assured. The outcome of the consultation process is printed in booklet form and every employee gets a copy. This is a wonderful tool to show everyone that their opinions matter and provides the moral high ground when rolling out the change programmes.

Call to Action: identify the people barriers to change

Meantime, managers at all levels are being appraised, identifying who is going to embrace what needs to be done, and who will get in the way of the necessary changes. It is my experience that in most programs some managers at the 3 levels need to be separated. For whatever reason, there are some that cannot or will not adapt to or accept the changes. Usually, I take on board a small team of consultants I know and trust to help me quickly identify those, and to educate and train people in the various programmes. It is absolutely essential that the people, particularly those at senior level who are not going to make it, are quickly dealt with. Otherwise, momentum will be slowed down and undermined by their presence. It must be emphasised that this is critical. The factory manager must be absolutely ruthless in driving this, because the greater good, perhaps the very survival of your plant, will be jeopardised.

Lean Toolkit - Making it happen

In parallel with this, we set about putting our Lean tool kit in place. There is no more effective way to obtain better results than through the "Lean journey." I like to have multiple programmes working in parallel. So, purchasing begins to work with key suppliers to get new working agreements in place. We establish Kanban to our warehouse and then to our factory floor, and for the internal movement of work in process. We train people in the art of Process

Mapping and get people working throughout the plant on generating improvements. We start to get people thinking about Value Streams, (factories within factories) and usually within 3 months have an exemplar Value Stream emerging, with others identified to work on later. Meantime, all of our plant reports are being redesigned so that every work centre has data that they can make sense of, enabling them to track their own performance. All data collected is analysed; if it is not adding value and being used to inform and improve it is eliminated. We also get our entire staff, including office and factory personnel, involved in redesigning their work areas through the 5S Process. This is an outstanding tool for maximum involvement of people and when implemented will transform how the entire factory looks.

Visible Results

With all of these programmes in play, credibility for the new programmes becomes well established. What has been promised is being delivered upon. The challenges facing the plant have been laid out backing it up with clear evidence. The workforce has been asked to answer the 4 key questions I referred to earlier, therefore taking their views on board. Those not delivering are being separated (workforces always know who those are and this further increase credibility). The factory is looking much better through the 5S programmes, and painting and refurbishment. Data that was not being used has been discontinued, reports from every department have been redesigned and now both inform and add real value. Communications processes are in place, and as much as possible this is done through Visual Management, with results and key messages posted throughout the plant. Space has been freed up in the warehouse and shop floor through the introduction of Kanban and implementation of 5S. Value Streams are being created to drive out waste with allocated people being decentralised so that they sit and work together. Usually after 3 months great excitement has been generated, with many programs already in place. Improved results are already visible. After 6 months a very substantial change is visible in both the atmosphere of the plant and its appearance, and of course its results. After 1 year it is truly spectacular, and then just keeps getting better as people become more familiar with the new way of working and with the various tools and techniques.

This is what works for me. I have achieved spectacular results by using this formula. The plant I was sent to close in the USA was declared a Benchmark site by Gillette within 2 years. The Ireland plant survived and still thrives as a Benchmark site. They continue to see off threats from Eastern Europe and Asia. In fact, they have just recently leased an adjoining factory. In my last assignment in Shanghai, I also left a Benchmark site behind. With this formula, applied without compromise, there is no failure, only spectacular success!

Why so many attempts at Change Management fail?

If what has been written so far is about how change works, what about failure? For me it is clear. If success is down to Leadership, then we are also responsible for failure. Any Leader that does not set out the context for change sufficiently and involve the workforce extensively at every level will simply not succeed. They must establish a communications process that is two-way with comments and questions listened and responded to. They have to provide the necessary education and resources to ensure successful implementation of the various programmes. Managers and employees that slow down or block the changes must be removed. It is within these critical areas that too many leaders compromise, and subsequently fail.

The Massachusetts Institute of Technology (MIT) has confirmed that over 60% of attempts at change fail. Personally, I think the figure is on the low side. From what I have seen over 3 decades it is much higher than that. I would have put it at around 80%. However, organisations are understandably reluctant to admit to failure, with very good reason. So, I suspect that always getting the real data is not so easy. Acknowledging failure means shattered careers, lost bonuses, and perhaps even a share slump. Divisions and departments close ranks and partial success or even downright failure is often celebrated as success, or the programme quietly dumped soon to be superseded by the next new flavour of the month. (Those readers working in large organisations will understand exactly what I mean!)

Over my career, I have been able to tap into people's natural instinct for survival, and natural desire for meaningful work. Workforces have the same needs all over the world, regardless of where they reside. They want to be respected, and to be acknowledged by having a voice that is listened and responded to. They want opportunity for education and development, and to be fairly remunerated so that they can provide security for themselves and their family. The wonder is why so many organisations struggle to see that, or to implement programmes putting them in place!

Finally, I conclude with one of my favourite descriptions of the organisation that I believe provides an essential basis for success!

The old rules of traditional hierarchical, high external control, top-down management are being dismantled they are simply not working any more. This has changed the role of manager from one who drives results and motivation from the outside into one who is servant leader, one who seeks to draw out, inspire and develop the best and highest within people from the inside out. '

- Stephen Covey

LIAM CASSIDY

Conclusion

I have chronicled my working life from when I entered the army at age sixteen, to AC Delco, Dunstable and London, back to Ireland and Packard Electric, then to Oral-B Laboratories, to Iowa, back to Ireland, and to the final assignment of my multinational career: Braun Oral-B in Shanghai. It's been quite a ride.

At every point in my career, I can look back and recall learning from experiences either directly or indirectly. But as important as it is to learn from your successes, it is equally important to learn from what was badly implemented so you do not repeat those mistakes.

Decades ago, I learned many things that I still consider to be "best practices," often from people with only a basic education. Today, I see them dressed up in language as if they were new. There are several effective, common-sense, manual processes that were used years and years ago that would be more effective today than some of the expensive systems organisations buy. The world is awash with data, churned out by expensive systems often costing many millions. Even worse, it isn't data that anyone trusts.

A couple of years ago, I was speaking at an event in Ohio. We had representatives from about 15 or so manufacturing organisations in the room. When I asked them, "Who here trusts their data?" not one person put up their hands. I was not surprised.

Many organisations are now rushing to embrace what many are calling the 4th Industrial Revolution, commonly referred to as Ind. 4.0. It is being touted by many as the solution to many of their problems.

Approach with caution!

I recently asked for opinions from four people in the business of providing systems to manufacturing, and I received four different understandings of Ind. 4.0. To be frank, I have been unable to get a definitive description of it anywhere. What I do know is that it envisages far more machine connectivity and a massive increase in data, and *this is what worries me.* Most organisations cannot handle the levels of data they have today. They simply do not have the levels of knowledge, discipline, and skill that are required to maintain it and react to it. So, what is the use of data if it is inaccurate, and you cannot use it?

As I have said, be cautious. Do not rush into it. If you feel you must, first embark on the Lean Journey. It's simply the best way to develop the skills, knowledge, and discipline to contemplate the road to Ind. 4.0. If it is possible to get there, the Lean way will do so. However, I still believe that Ind. 4.0 should be approached with great caution with a clear view of the resources required not just to implement such a move, but to *maintain* it.

I have seen more automated data systems on factory floors than I can recall and have yet to see one that delivered on what was promised. There are so many things going on in busy factories, so many variables that exist, that to maintain accurate databases is a real challenge and one that few can meet. Moreover, automated systems tend to separate those actually doing the work from the data, because they don't control it. As I have explained previously, when employees are responsible for recording their own data, it keeps them emotionally engaged with what they do and leads to far better results.

My message is to keep things simple. Do not seek complicated systems or automate simple processes just for the sake of it, especially if everything is already working well.

As I've said, despite being an advocate for it, Lean has a high failure rate. A couple of years ago, the Lean Enterprise Institute's research confirmed it was over 90%. This came as no surprise to me, but why is this so?

Because organisations often seek instant solutions to difficult issues, and they are not prepared to devote the kind of time and support required for a Lean Program to be successful. Committing to radical change takes full commitment. When I tell a leadership team or CEO what is required, they most often say yes without understanding what that means.

What it means for the factory is that leadership must stop fixing the same issues every day, and instead must work towards creating an environment where they can focus on the future. Leadership is about putting strong processes in place with empowered and well-trained people running them. Their job is to ensure that repetitive issues are fixed at the root, not stitched together so they will reoccur later. The organisation must invest in their entire workforce so that they are continuously developing the skills necessary to deal with today and tomorrow.

It is also the site leader's job to identify those individuals on his/her team who are not going to contribute sufficiently, and then separate them from the operation. There should be no ambiguity about this, but sadly organisations will often put their operation at risk by carrying ineffective managers. This is a failure of top-level leadership and should not be seen in any other way. I preach constantly to my clients and students that leadership must give precedence to the greater good, and not to misplaced loyalty to a colleague or subordinate.

Throughout this book I have shown you how, after almost five decades of experience in manufacturing, I believe Lean Thinking and Practices are still by far the most effective way to run your operation. No other approach will come even close to providing continuously improving results, and it energises and involves employees at all levels. However, I have also laced this book with common-sense lessons and applications that I learned long before Lean was ever developed. Much of it is not new, but much of it bears repeating.

Sadly, consultancy work in manufacturing today is populated by many who do not know enough about what they are doing, or they do but will not put their project at risk by insisting their clients follow through with tough decisions and radical change. Too often, they defer to short-term revenue flow for their consultancy group instead of what is truly best for the client. They set their clients up to fail, and it is a modern tragedy in the manufacturing world today.

If you read this book because you are worried about your site and have decided to seek third-party help, exercise caution. There are many offering their services, but only a fraction leaving sustainable and successful change behind them.

There is the right kind of help out there—just take some time and seek it out. Because *now* you know what to look for!

Connect with the Author

Thank you for reading *Make Your Factory Great & Keep It That Way!*

If you enjoyed this book and would recommend it to your industry colleagues and friends, please do not hesitate to leave a review on Amazon, Barnes & Noble, or wherever you purchase your books.

If you wish to speak with the author, Liam Cassidy is available for:

> ➢ Leadership coaching and mentoring (Teams and individuals)
> ➢ Manufacturing & Supply Chain strategies
> ➢ Lean Transformations
> ➢ Speaking engagements

You may reach him at: liam.cassidy@altixconsulting.com

Altix Consulting are the middle market international industrial champions' management consulting partner. We provide business strategy, technology and innovation, and operational excellence support, in the world of advanced manufacturing and international supply chain.

If you wish to work with Altix Consulting, please visit: www.altixconsulting.com

Glossary

5S

A method used to organise the work area. Considered the foundation of Lean Practices.

- *Sort*: eliminate that which is not needed
- *Straighten:* organise remaining items
- *Shine:* clean and inspect work area
- *Standardise:* write standards for above
- *Sustain:* regularly apply the standards

Autonomous Maintenance

A method from Total Productive Maintenance (TPM) for engaging operators to carry out basic maintenance activity (such as cleaning, lubrication, and inspection activity).

Backflushing

Backflush accounting is when you wait until the manufacture of a product has been completed, and then record all of the related issuances of inventory from stock that were required to create the product.

Business Groups

Later called Value Stream Organisations. Similar products brought together and independently managed by multidisciplinary teams that are transferred from central departments. Leads to more efficient flow and adds value for customers.

Continuous Flow

Manufacturing where work-in-process smoothly flows through production with minimal (or no) buffers between steps of the manufacturing process.

ERP
Enterprise Resource Planning

FIFO
First in first out

Gemba (The Real Place)
A philosophy that reminds us to get out of our offices and spend time on the plant floor—the place where real action occurs.

Heijunka (Mixed Modelling)
A form of production scheduling that purposely manufactures in much smaller batches by sequencing (mixing) product variants within the same process.

Just-In-Time (JIT)
Pulling parts through production based on customer demand instead of pushing parts through production based on projected demand. It relies on many Lean Tools, such as Continuous Flow, Heijunka, Kanban, Standardised Work, and Takt Time.

Kaizen (Continuous Improvement)
A strategy where employees work together proactively to achieve regular, incremental improvements in the manufacturing process.

Kanban (Pull System)
A method of regulating the flow of goods both within the factory and with outside suppliers and customers. Based on automatic replenishment through signal cards that indicate when more goods are needed.

Muda

Refers to anything in the manufacturing process that does not add value from the customer's perspective.

Overall Equipment Effectiveness (OEE)

A framework for measuring productivity loss for a given manufacturing process. Three categories of loss are tracked:

- Availability (i.e., downtime)
- Performance (i.e., slow cycles)
- Quality (i.e., rejects)

PDCA (Plan, Do, Check, Act)

An iterative methodology for implementing improvements:

- *Plan:* establish plan and expected results
- *Do:* implement plan
- *Check:* verify expected results achieved
- *Act:* review and assess; do it again

Single Minute Exchange of Die (SMED)

Reduces setup (changeover) time to less than 10 minutes. Techniques include:

- Convert setup steps to be external (performed while the process is running)
- Simplify internal setup (i.e., replace bolts with knobs and levers)
- Eliminate non-essential operations
- Create standardised work instructions

SMART Goals

Goals that are: Specific, Measurable, Attainable, Relevant, and Time-Specific.

Standardised Work
Documented procedures for manufacturing that capture best practices (including the time to complete each task). It must be "living" documentation that is easy to change.

STAR Products
High regular volume demand.

Takt Time
The pace of production (i.e., manufacturing one piece every 30 seconds) that aligns production with customer demand. It is calculated as Planned Production Time / Customer Demand.

Total Productive Maintenance (TPM)
A holistic approach to maintenance that focuses on proactive and preventative maintenance to maximise the operational time of equipment. TPM blurs the distinction between maintenance and production by placing a strong emphasis on empowering operators to help maintain their equipment.

Value Stream Mapping
A tool used to visually map the flow of production. It shows the current and future state of processes in a way that highlights opportunities for improvement.

Value Stream Organisation (VSO)
All necessary functional representatives grouped and working together to maximise the results of the value stream. Mostly found within Lean environments.

CPSIA information can be obtained
at www.ICGtesting.com
Printed in the USA
BVHW011644140422
634332BV00010B/390

9 781399 918022